Microstrip Multipliers Realization

Kamaljeet Singh

ATLANTIC
PUBLISHERS & DISTRIBUTORS (P) LTD

Published by

ATLANTIC

PUBLISHERS & DISTRIBUTORS (P) LTD

7/22, Ansari Road, Darya Ganj, New Delhi-110002
Phones : +91-11-40775252, 40775214, 23273880, 23275880
Fax: +91-11-23285873
Web: www.atlanticbooks.com
E-mail: orders@atlanticbooks.com

Printed & bound in India by Atlantic Print Services

We make a living by what we get,
We make a life by what we give.

—Winston Churchill

Preface

Frequency multiplier designs use a variety of nonlinear devices and topologies to achieve excitation of harmonics. Frequency multiplications using diode, transistor, and mixer devices rely on the principle of non-linearity which distorts signal waveform and generates harmonics. Frequency multipliers are employed in high-frequency circuits where stability is the key requirement.

Signal generation from higher frequencies is realized from low frequency sources utilizing multipliers resulting in better noise performance. Low frequency sources employing crystal, manufactured with resonances beyond 100 MHz, are fragile, hard to manufacture, and expensive. Attaining higher frequencies are achievable by the implementation of the multiplier chain. Traditionally, Gunn diode or IMPATT oscillators are employed but lately replaced by solid state devices as former were expensive, noisy, and bulkier. MIC based approach employing diodes or transistor replaced due to lightweight, reliability, inexpensive, and compactness. The active devices can be FET such as MESFET, HEMT, or diodes such as SRD can be employed for multiplication. Basic methods of frequency generation are direct signal generation, sub-harmonic mixer, up-converter mixer, and frequency multipliers.

Direct signal generation for high frequency generation utilizing oscillator tuned to the desired frequency such as dielectric resonator oscillator (DRO). The main challenge in this approach lies in achieving stability and avoidance of distortion and noise. Alternatively, multiplier based approach having low frequency oscillator followed by multiplier to achieve high frequency at the desired frequency results in better noise and stability in performance.

FET based multipliers can be used up to 3-4 times of fundamental frequency whereas diode based multipliers are employed for higher order. Diode based multipliers are insensitive to ESD failures but still preferred to be employed with the limitation of higher losses. Apart from the above, mixers are also attractive options but employ two simultaneous sources for the generation of various components. Microwave integrated circuits beside MMIC are the potential and widely employed topologies. Fundamental principles in both the topologies are same such as design and e-m simulation but various other aspects such as fabrication, modelling, and packaging are having different concepts. Moresoever, MMIC basically deals with only transistor based topologies realized only on GaAs.

The basic component in multiplier topology is the non-linear device followed by filter for the selection of the desired frequency. The basic operations of multipliers consist of:

1. Non-linear device for generating higher harmonics,
2. Unwanted signal filtration by employing suitable filters,
3. Undesired harmonics or corresponding currents to be shorted, and
4. Proper matching to achieve desired performances.

Diode based multipliers are robust and in-sensitive to the ESD related failures. Higher order multipliers require higher drive power with degraded efficiency. The diodes preferably employed are varactors and SRD.

Main considerations in the multiplier realization cover impedance matching and biasing arrangement apart from output filtering to achieve the desired results. Multipliers are basically employed in the transponder chain consisting of both receiver and transmitter. In the super-heterodyne receiver, the front end chain consisting of 1st and 2nd stage of mixer are fed with LO frequency which is generated by TCXO and multiplied to achieve the desired IF frequency.

Transceiver plays an important role in the communication system which provides the basis of the overall system. Multiplier in the transmitter chain plays an important role as the synthesized carrier signal generated with the TCXO is fed to the diode followed by amplifier. The phase modulated signal is doubled using BJT based multiplier. Further to achieve operations at Ku-band, C-band/S-band frequency generation is followed by frequency doublers and triplers using the FET devices. Modelling of semi-conductor device is having two approaches—device physics and measurement based approach. Device based approach translates physical structure and operating mechanism of the device into circuit mode whereas parameters of the lumped circuit model are obtained in the measurement based approach and in the present book later approach is adopted to simplify the circuit realization. Planar topology primarily MIC technique employed to realize multipliers in MIC and MMIC topologies are preferred. Compared to MMIC, MIC based topology is having ease of implementation. Present book details the MIC based topology for multiplier realization with implementation approach.

The first chapter deals with the RF fundamentals, Maxwell's equation, and basics of transmission lines. Simpler mathematical equations and illustrations have been provided so as to make it simpler to understand subsequent chapters. Microstrip features and MIC fabrications details have also been covered in this chapter.

The second chapter briefly details various multiplier topologies and their comparative analysis. The fabrication aspects apart from dedicated section on modelling are also incorporated in this chapter.

The third chapter deals with the diode based multiplier. C-Band frequency quintripler (×5) in microstrip configuration using step recovery diodes is demonstrated. SRD diode has been practically modelled and fitted in SPICE model before implementing in the circuit simulation tool and analyzed using harmonic balance techniques. The design is based on self-bias topology and repeatability is assured by eliminating tuning elements. Main feature is the planar implementation of SRD multiplier using CAD tools giving close match with simulated values.

The fourth chapter presents multipliers using HEMT which uses harmonic loaded filter. HEMT is modelled using quasi-static approach assuming the time-invariant lumped elements which depend only on the DC bias and amplitude of the terminal dynamic voltages. The drain-source current is represented by Materka model which is fine tuned to have close match between simulated and measured AC and DC data. At two different Ku-band frequencies, tripler circuits are simulated and fabricated using the derived model.

The fifth chapter presents implementation of single-ended mixer using diode which is modelled taking both diffusion and depletion capacitances into account. Suppression of harmonics of the local oscillator frequencies and inter-modulation products are achieved by placement of resonant stubs at these frequencies. A novel approach for the low pass filter at IF is adopted based on planar microstrip elements to reduce the area required for the filter.

Finally, the sixth chapter presents system level approach showing design of various components and overall system level simulation as well as link budget calculation. Further, latest techniques are also presented.

The present book deals with the MIC based multiple topologies which covers both diodes and transistors based topologies. Higher efficiency and lower loss or gain are the basic advantage associated with transistor based topology.

This book is intended for the researchers, students, and keen enthusiasts in the concerned field to imbibe the various aspects of multipliers with emphasis on implementation aspects. It covers practical approaches related with the design, fabrication, and characterization aspects with practical examples. As the discussed area has the potential for further extensive research, so more details can be found in the attached references. The book has been written in simple language starting with fundamentals and showing easier implementation methodologies apart from modelling aspects.

My sincere appreciation to colleagues and seniors at URSC apart from my family members for the wholeheartedly support in writing this book. Also, my sincere thank to all the sources which were referred during the writing of this book and my heartfelt apologies due to inadvertently omission of the same.

My appreciation to the publisher for the effort to show the book light of the day. I wish readers will be benefitted from the book and their constructive suggestions for further improvement of the presented work are welcome, as Magnus N Amudi stated, "A good education must expose knowledge gap and increases hunger for further inquires".

Dr. Kamaljeet Singh

Acknowledgements

My sincere gratitude to Director, URSC, for all the encouragement and promoting creativity as well as effective R&D culture in spite of multiple activities running concurrently in the organization.

I am indebted to the Chairman, ISRO, for giving me the opportunity and full access to resources without which it was impossible to finish this work. I am always thankful to the ISRO culture for providing the great working environment with competent and inspiring leadership. I would also like to thank my well-wishers, near and distant, who have kept me in their thoughts and prayers.

My heartfelt gratitude to Dr. Surendra Pal and Sh. K. Nagachenchaiah for their unflinching support, encouragement, and shaping my carrier at ISRO with motivation and guidance. Their supportive behaviour and guidance attitude are truly awesome and inspiring. The culture prevalent at ISRO motivates and dedication of people to quality and ethics of research makes unforgettable mark.

My sincere thanks to Prof. Deepak Bhatnagar, Late Sh. R.M. Mishra, Dr. V.K. Lakshmeeshah, Sh. R. Ramsubramanian, Prof. Sandeep Sancheti, Madam Selvi, Dr. S.V. Sharma, and specially to Sh. A.V. Nirmal, for his moral support and contribution towards my development. I thank them from the bottom of my heart. I am also thankful to various people who all compelled me to give back to the society in various form. Thanks are also due to my friends and colleagues at URSC and various groups for paving the way of learning each day.

Last but not least, I am most thankful to my family for their boundless love and support, who never failed to believe in me.

I learned this subject from researchers and authors who are too many to include in this short space, but I gratefully acknowledge their contributions in providing me new insight into this topic.

I express my thankfulness to THE ALMIGHTY who has given me inner strength in completing this book within short span of time. This book is dedicated to students and researchers depicting simple implementation of the multiplier topologies in MIC which is integral part of communication system.

Dr. Kamaljeet Singh

List of Abbreviations

BPF	Band pass filter
BSF	Band stop filter
BW	Band width
CAD	Computer aided design
CF	Center frequency
CL	Conversion loss
CMOS	Complimentary metal-oxide structures
CNR	Carrier to noise ratio
CPW	Coplanar wave-guide
DBR	Dual band resonator
DR	Dielectric resonator
DRO	Dielectric resonator oscillator
EBG	Electronic band gap
HEMT	High electron mobility transistor
HRSi	High resistivity silicon
IF	Intermediate frequency
IM	Inter-modulation
LPF	Low pass filter
LRS	Low resistivity wafer
MCM	Multi chip module
MMIC	Monolithic microwave integrated circuit
MIC	Microwave integrated circuits
MEMS	Micro electro-mechanical systems
JEDEC	Joint Electronic Devices Engineering Council
OCXO	Oven controlled crystal oscillator

PCI	Peripheral component interface
PCL	Parallel coupled line
PLL	Programming phase locked loop
PFD	Phase frequency divider
SIR	Stepped impedance resonator
SoC	System on chip
SPICE	Simulation program with integrated circuit emphasis
SRD	Step recovery diode
SRF	Series resonant frequency
TCXO	Temperature compensated crystal oscillator
UIR	Uniform impedance resonator
UWB	Ultra wide band
VCO	Voltage controlled oscillator

Physical Constants

Impedance of free space, η_o	377Ω
Velocity of light in free space, c	$3.0 \times 10^8 \times$ m/s
Charge of electron, q_e	1.602×10^{-19} C
Permittivity of free space, ε_o	8.8542×10^{-12} F/m
Velocity of light in free space, c	2.997925×10^8 m/s
Charge of electron, q_e	1.60210×10^{-19} C
Mass of electron, m_e	9.1091×10^{-31} kg
Boltzmann's constant, k	1.38×10^{-23} J/K
Planck's constant, h	6.6256×10^{-34} J·s

General Formulas

General Formulae:

Speed = distance/time

Velocity = displacement/time

Density = mass/volume

Acceleration = change in velocity/time

Pressure = force/area

Power = work/time

Impulse = force × time

Momentum = mass × velocity

Surface tension = force/length

Frequency = 1/T

Definitions

Microwave: Alternating current signal with frequencies ranging between 10^6 Hz to 10^9 Hz.

Radio frequency: Alternating current signal with frequencies ranging between 10^4 Hz to 10^{12} Hz.

Filter: Two-port network for controlling frequency response by passing or attenuating signals.

Intrinsic impedance η: This impedance solely depend on the material parameters of the medium.

Wave impedance: This impedance is associated with the particular type of wave and depends on the type of line, material and operating frequency.

Characteristic impedance: Impedance associated with matched line

Sensor: It is a device whose output can be quantified and changes with one or more physical phenomena and the output information can be used for process monitoring and control.

Transducer: It is a device that converts one form of energy into another form. The measurement of physical variables associated with the resulting form of energy allows estimation of the physical variables associated with the input energy.

Detector It is a device indicating presence, absence, or change of the signal qualitatively, either as a binary signal or as a low resolution signal with several states.

Frequency divider: It is an electronic device that produces a signal at 1/N of the input frequency where N is an integer.

Wavelength: Electromagnetic radiation consists of electromagnetic waves created due to periodic change of electric field or magnetic field producing different wavelength of spectrum represented as $\lambda = c/f$.

Spatial wavelength of the periodic interdigital structure is the distance between the center lines of the adjacent fingers belonging to the same electrode.

Permittivity is the material property describing the electric polarization.

Permeability is the material property describing the magnetic polarization.

Signal Bandwidth measure the frequency spectrum occupied by the signal.

Filter bandwidth measures the frequency range which can be passed by the filter.

Channel bandwidth measures the overall bandwidth of the transmission channel.

Conductor is a material which conducts electricity or /and heat. Copper and aluminium are conductors whereas rubber, PVC, mica, and plastics are insulators.

Printed circuit is an electronic circuit that is built on an insulated board without the need for conducting wires.

VOP (velocity of propagation): It is determined as the speed at which signal propagates and defined in terms of percentage of the free space value of the speed of light.

Total Harmonic Distortion: Ratio of the sum of the powers of all harmonic components to the power of the fundamental frequency.

P1-dB (1-dB compression point): Output level at which amplifier's gain is 1-dB less than the small signal gain.

Dynamic range(DR): The ratio of the signal power that causes distortion power (for single frequency component) equal to the noise floor and the minimum detectable signal.

$$DR = \frac{2}{3}[P_{IP}(dBm) - N(dBm]$$

(a) SFDR: Ratio of fundamental signal to the strongest spurious signal at the output.

(b) LDR: Ration between the largest and smallest values that certain quantity can assume.

Dynamic range of ADC= 6.021 N+1.763 dB

N= no of bits

So, for 12 bit it will be 74 dB

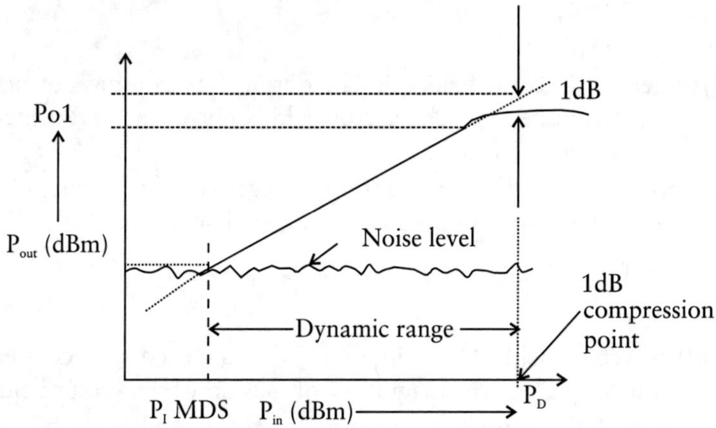

Characteristics of a non-linear circuit

Linearity: Output power is proportional to input level.

Pout (dBm)= Pin(dBm)+ G(dB)

Intermodulation products (IMD): Amplitude modulation of signals containing two or more different frequencies caused by non-linearities or time variance in the system. It creates spurious frequency components.

Fundamental frequency: Lowest frequency of a periodic waveform.

Total harmonic distortion: It is the ratio of the sum of the powers of all harmonic components to the power of the fundamental frequency.

Bolometer: Device changes resistance with the absorption of power and having positive temperature coefficient of resistivity whereas thermistors are having negative temperature coefficient of resistivity.

Thermocouple: It generates d-c voltage with the absorption of power.

Sensitor: It is resistor having the exponentially dependence on temperature with positive temperature coefficient.

Voltage: The voltage difference between two points in the vector filed equals to the line integral of the intensity of the vector field and its direction is opposite to the direction of the field.

$$\phi_p - \phi_q = -\int_q^p \overline{A}.\overline{dr}$$

Divergence of vector field(∇.): The output flux coming out per unit volume from the vector field is known as divergence of the vector filed.

Gauss divergence theorem: Surface integral of the vector field equals to the volume integral of its divergence.

$$\int_S \overline{A}.\overline{da} = \oint (\nabla \times \overline{A}) dV$$

Curl of vector field (Δ×): Line integral and closed surface area ratio equals to the magnitude of the curl of the vector field.

Stokes curl theorem: Line integral for the closed surface and corresponding area in the vector field equals to the surface integral

$$\int_S \overline{A}.\overline{da} = \oint_V (\Delta.\overline{A}) da$$

Linear medium: The medium in which permittivity value changes linearly with the electric field.

Homogeneous medium: The medium in which the permittivity associated with the material is not changed and equals at all points.

Anisotropic medium: The reference changes with the space coordinates than medium is considered to be anisotropic.

Isotropic medium: Electric filed intensity and electric flux density are independent of directions.

Antenna gain (G) is defined as the ratio of the power produced by the antenna from a far-field source on the antenna's beam axis to the power produced by a hypothetical lossless isotropic antenna. This ratio is expressed in decibels-isotropic (dBi).

Power radiated (in Watts) = $P(\Theta,\Phi)\ \Omega_A$

Ω_A (in Watts) $= \int_0^{2\pi} \int_0^{\pi} P(\Theta, \Phi) d\Omega$

$d\Omega = \sin\Theta\ d\Theta\ d\Phi$

Contents

Chapter 1

RF Basics

1.1 Introduction

Electro-magnetic wave composed of electric and magnetic fields. They are generated from a point source and spreads in all directions as spherical waves. The electro-magnetic waves possess energy and the main classifications are radio-frequency, microwave, infrared, visible light, ultra-violet, X-rays and gamma rays. These wave travel with the velocity of light ($c=3\times10^8$ m/s). The main difference of wavelength among them relates with the propagation of electro-magnetic energy. The wave propagation is equivalent to the water waves as it propagates near to the surface as well as radiates in the space due to reflection at various angles. The radio waves are having two components—crest and trough and its propagation is effected by the medium which can be either transmission line or air. The electromagnetic wave are transverse in nature where electric field, magnetic field and wave propagation are transverse to each other. The fundamental rules for high frequency and microwaves are same but at high frequency due to field specific behaviour, the explanations in terms of voltages and currents are difficult [1]. Another important aspect of distributed transmission line parameter at different points. The mathematical description can be represented as:

$$\lambda = c/f \text{ and } E = \frac{hc}{\lambda} \qquad \text{... (1.1)}$$

where c is the speed of the light in vacuum, f is the frequency (Hz), λ is the wavelength, h is the Plank's constant.

So according to equation (1.1) due to low and high frequency; the wavelength differs with the change of frequency. The waveguide width at 10 GHz is 30 cm (vacuum propagation) and turned out to be 300 cm for 1 GHz. So it indicates that ten times increase in the frequency results in ten times reduction of waveguide dimensions.

The effect of permittivity ε, if incorporated, the speed of the e m wave becomes inversely proportional to the square root of permittivity and can be represented as:

$$\lambda_g = \frac{\lambda_0}{\sqrt{\varepsilon_{eff}}}; E = \frac{hc}{\lambda_g} \quad \dots (1.2)$$

λ_g: guided wavelength, λ_g: wavelength in air/vacuum, ε_{eff}: effective permittivity in the medium

Table 1.1: Relative permittivity of the material

Material	Relative Permittivity
Vacuum	1
Air	1.0006
Teflon	2.1
Alumina	9-11

1.1.1 Advantages of High Frequency Signals

Enhanced bandwidth: At higher frequencies, the achievable bandwidth enhances resulting in transmission of more data, voice, video streaming, etc. Taking 1% bandwidth of 2 kHz results in 20 Hz whereas 2 GHz the 1% results in the BW of 20 MHz resulting in 1 kHz increase.

Antenna Gain: The gain of antenna is dependent on the wave-length and aperture of the antenna. So at higher frequencies and larger aperture achieving higher gain is possible.

Thermal effect: Higher frequencies is associated with the thermal effect and same is utilized in the microwave oven

Line of sight propagation: Line of sight propagation at high frequencies which is not diverted (differentiated) due to ionosphere and become useful in wireless communication.

Other applications: Various molecular, atomic, nuclear resonances at higher frequencies are having applications in remote sensing.

Table 1.2: Microwave frequency bands

Band	Frequency Range	~Wave-length
L	1-2	15 cm
S	2-4	7.5 cm
C	4-8	3.75 cm
X	8-12	25 mm
Ku	12-18	16.6 mm
K	18-26	10.5 mm
Ka	26-40	7.5 mm
U	40-60	5 mm
V	60-80	2.5 mm
W	70-110	1 mm

The main challenge at high frequency circuits related with circuit analysis because at low frequency circuit element can be analyzed using Kirchoff and Norton theorem but at high frequency the circuits elements are distributed. The voltage and current at each point of distributed circuit changes and necessitate field assumptions which can be solved using Maxwell's equation [2]. At high frequencies losses are shown in terms of dB or dBm (Ann-1). VSWR (voltage standing wave ratio) of 1.2 or less means energy transfer of 99.8%. In case of –20 dB loss, 99% got attenuated and for –30 dB loss, 99.9% of radio frequency got attenuated. VNA (vector network analyzer) used for the measurement of losses.

1.1.2 Characteristics of Electro-magnetic Waves

Radiation is the main characteristics associated with the electro-magnetic waves which is considered unwarranted for circuits but desirable for antennas. Various phenomena such as reflection, refraction, diffraction, interferences are associated with the em waves resulting in cross pole and side lobes in the antenna. Linear passive topologies act as transmitter as well as receiver resulting in the single antenna usage such as Duplexer or Diplexer. Also radio waves get attenuated while propagating and its value inversely changes with the distance (R) which can be shown as Attenuation $\alpha \frac{1}{R^2}$.

Table 1.3: Radio frequency bands and applications

Band	Frequency Range	Applications
Very low frequency	10^3-10^4	AM
Low frequency	10^4-10^5	Frequency modulated
Medium frequency	10^5-10^6	Low wave radio
High frequency	10^6-10^7	Frequency modulated radio
Very high frequency	10^7-10^8	High frequency TV
Ultra high frequency	10^8-10^9	
Super high frequency	10^9-10^{10}	Satellite communication
Extremely high frequency	10^{10}-10^{11}	Radar, microwave transmission
Infrared	10^{11}-10^{14}	Photography
Visible light	10^{14}-10^{15}	Vision
Ultra violet	10^{15}-10^{16}	Parasitics removal
X-rays	10^{16}-10^{19}	Medical, Inspection
Gamma rays	10^{19}-10^{21}	Cancer therapy
Cosmic rays	$>10^{21}$	Astronomy

The propagation of radio waves in the medium are of two types—1. Mechanical waves and 2. Electro-magnetic waves.

Mechanical waves can be considered as disturbance which needs medium for propagation from one place to another. These waves are also known as elastic waves and are classified as—transverse and longitudinal waves. At high frequencies charged particles, ionized gases, transition of atomic electrons from higher state to lower and sinusoidal current flow at high frequencies results in specific waves different from mechanical waves which can travel in vacuum also. The analysis of nature and shape of these waves can be represented by the differential equations. The solution of these equations shows that these equations are equivalent to electric and magnetic wave equation whose velocity is equivalent to the velocity of light. The main characteristics of these waves are:

1. The velocity of electro-magnetic waves are equivalent to the light velocity
2. The component of electric and magnetic fields of electromagnetic waves along the direction of wave propagation is zero which indicates em waves are planar in nature.
3. The electric and magnetic fields in the planar electromagnetic waves travel perpendicular with the same phase.

Both energy and momentum are propagated with the em waves from one place to another. The change in momentum due to attenuation or refraction due to impinging of the waves on to the surface resulting in the pressure which is known as radiation pressure. Radio waves, microwaves, infrared rays, light waves, X-rays, Gamma rays, cosmic rays are various types of electro-magnetic waves.

Electromagnetic waves are due to combination of electric and magnetic fields [3] where electrostatic fields generated from static electric charges whereas magnetic fields originated from static magnetic charges or magnetic current. Time dependent phenomena responsible for the origin of electromagnetic waves which can be represented by four fundamental equations termed as Maxwell's equations. Basically, the equations are derived from the Gauss, Faraday's and Ampere rules. The equations given by Maxwell in 1862 are detailed in subsequent section.

1.2 Maxwell Equations

Maxwell's equations are the basic in electromagnetic theory and basically are categorized in four types which are governed by the fundamental laws. At higher frequencies the circuit concept is changed into filed concept as shown in Table 1.4.

Table 1.4: Equivalence between filed and circuit concept

Circuit Concept	Filed Concept
Voltage v	Electric intensity
Current i	Magnetic Intensity
Charge q	Electric flux density

Circuit Concept	Filed Concept
Energy Flow *vi*	E × H
Capacitive Energy $E_c = \frac{1}{2}CV^2$	Electric energy $E_e = \frac{1}{2}\varepsilon E^2$

Maxwell First Equation: The first equation is based on Gauss law which states that electric flux (Φ_E) from closed surface and total closed charges (Σq) are related as below:

$$\Phi_E = \frac{\sum q}{\varepsilon} \quad \text{... (1.3)}$$

The above equation can be represented in integral form as

$$\oint_S E.da = \frac{1}{\varepsilon}\int_V \rho dV \quad \text{... (1.4)}$$

$\Phi_E \rightarrow$ output flux from the closed surface

$\Sigma q \rightarrow$ Closed charge

Using the Gauss divergence theorem, the above equation can be written as

$$\nabla.E = \rho/\varepsilon \quad \text{... (1.5)}$$

$\rho \rightarrow$ charge volume density

Since $D = \varepsilon\, E$, so equation (1.5) can be written as (1.5)

$$\nabla.D = \rho_V \quad \text{... (1.6)}$$

Maxwell Second Equation: Maxwell second equation is based on the Gauss law for magnetic field. As per Gauss law for magnetic field, magnetic lines in the magnetic field are always closed which can be interpret as magnetic force lines are not having source and sink. Magnetic flux lines are without source and sink. The total flux coming out from the enclosed surface (net input-output) is zero which can be represented by the following

$$\Phi_B = 0 \quad \text{... (1.7)}$$

Using Gauss divergence theorem in integral form

$$\oint_S B.da = \frac{1}{\varepsilon}\int_V (\nabla.B)dV$$

Resulting in $\nabla.B = 0$ ... (1.8)

Equation (1.8) is the differential form of Maxwell's second equation.

Maxwell Third Equation: This equation is based on the electro-magnetic inductance rule given by Faraday. It shows the relation between the relative change in the magnetic field with time and its linkage with the electric field. According to the definition, the induced electro-magnetic force is equal to the change of rate of magnetic flux where the induced emf direction is such that it opposes the rate of change of magnetic flux as shown below

So magnetic flux $\Phi_M = -\frac{\partial}{\partial t}\int_S B.da$

Using Stokes theorem

$$\oint_c E.dl = \int_S (\nabla \times E).da$$

$$\nabla \times E = -\frac{\partial B}{\partial t} \qquad \text{... (1.9)}$$

Negative sign is associated with the self-inductance and introduced by Lenz's rule.

Maxwell Fourth Equation: Maxwell fourth equation deals with the current and associated magnetic flux. It can be found out using Ampere's circuital rule. As per Ampere's rule, the line integral of the magnetic field for the closed circuit equals to the total current and can be represented as

$$\oint_c H.dl = I$$

$$\oint_c B.dr = \int_S (\nabla \times B).da$$

The equation can be written in differential form as below:

$\nabla \times B = \mu J$; where μ (medium permeability).

The above equation related the relation between the constant current with the magnetic field. The relation for the change with respect to time can be shown as:

$$\nabla \times B = \mu J + \frac{\partial D}{\partial t} \qquad \text{... (1.10)}$$

$\mu = \mu_0 (1 + X)$; $X = I/H$ = magnetic intensity/magnetic field where $\mu = \mu_0 (1 + X)$ is the magnetic permeability of the material. In case of vacuum:

$\mu = \mu_0$, the relative permeability $\mu_r = \mu/\mu_0 = 1 + X$ and X is known as magnetic susceptibility

The above equation is the combination of two currents: conduction and displacement current

B	–	Magnetic flux density	(Wb/m^2)
H	–	Magnetic field intensity	(A/m)
D	–	Electric flux density	(C/m^2)
E	–	Electric filed intensity	(V/m)
J	–	Current density	(A/m^2)
X	–	Curl	
∇	–	Divergence	
da	–	Infinitesimal area	
dV	–	Infinitesimal volume	
ρ_V	–	electric charge density	(c/m^2)

Maxwell equation changes according to the medium. Medium can be open space, insulator or conductive and accordingly conductivity, permittivity and permeability changes which is shown in Table 1.5.

Table 1.5: Maxwell equation in various medium

Medium	Open space	Linear medium	Isotropic medium
First Equation	$\nabla.E = 0$	$\nabla.D = \rho$	$\nabla.E = \rho/\varepsilon$
Second Equation	$\nabla.B = 0$	$\nabla.B = 0$	$\nabla.B = 0$
Third Equation	$\nabla \times E = -\frac{\partial B}{\partial t}$	$\nabla \times E = -\mu\frac{\partial H}{\partial t}$	$\nabla\times E = -J\omega B$
Fourth Equation	$\nabla \times H = \frac{\partial D}{\partial t}$	$\nabla \times H = \varepsilon\frac{\partial E}{\partial t} + J$	$\nabla\times H = J\omega\varepsilon E + J$

Complex electromagnetic equations based on FEM techniques are also based on Maxwell's equations which are solved employing the boundary conditions.

Concept of Displacement Current: Maxwell equation $\nabla \times B = \mu J + \frac{\partial D}{\partial t}$ indicates change in electric field with time results in magnetic field so the changing electric field is equivalent to the current. This current known as displacement current and mathematically described by the parallel plate capacitance connected with the battery. So the current flow in the circuit, even though existence of discontinuity between the plates. This can be described as :

$$D = \varepsilon_0 E + P = \varepsilon_0 \varepsilon_r E$$

Electric field between capacitance plates E = σ_f / ε'; where σ_f open charge density c/m^2

The current between capacitance plates

I = dq/dt and the current density J = dσ/dt

J = dD/dt = jωD (isotropic electric field)

So current density is equivalent to the change of rate of displacement vector. This is equal to the current density in the outer circuit which is known as displacement vector.

The displacement current density results in current continuity in the overall circuit.

$$\nabla \times H = J = J + J = j\omega D = j\omega(\varepsilon' - j\varepsilon' - j - j\frac{\sigma}{\omega}) = j\omega\varepsilon^* E$$

For a dielectric

$$\tan\delta = \frac{\omega\varepsilon'' + \sigma}{\omega\varepsilon'} = \tan\delta_d + \tan\delta_c$$

The currents are the combination of two current components conduction and displacement.

1.3 Transmission Lines

Transmission lines are the medium used to propagate the energy of electric and magnetic waves. The main purpose of the transmission lines is to transmit radio frequency from transmitter to antenna and from antenna to the receiver. It can be used as circuit element also such as impedance (resistance), inductance, capacitance, resonant circuits, filters, impedance matching, etc.

at radio frequencies. At higher frequencies, transmission line play important role as the phase at each point varies . This effect along with equal distribution of capacitance and inductance at each point of transmission line makes the distributed line parameters. Subsequently the constant associated with transmission line are dependent on the unit length [4]. The equivalence is shown as below:

The primary constant associated with the transmission lines are resistance (R), inductance (L), capacitance (C), conductance (G):

Distributed Resistance: The resistance of conductor are distributed equally across the line length and represented as (R).

Distributed Inductance: The inductance is connected in series with the resistance magnetic field introduces inductance which is equally distributed across transmission lines and represented as (R).

Distributed Capacitance: Parallel capacitance generated due to parallel plate phenomena which is associated with the parallel conductors and represented as shunt capacitance (C).

Distributed Conductance: It is associated with the distributed current due to lossy dielectric. The effect is measured in terms of conductance and represented as parallel resistance. The dissipate current is represented as (G).

The secondary constant can be represented as:

$$\gamma = \sqrt{(R + j\omega L)(G + j\omega C)}\sqrt{(R + j\omega L)(G + j\omega C)} = \alpha + j\beta$$

$$\beta = \omega\sqrt{LC}$$

$$Z_0 = \sqrt{\frac{R + j\omega L}{G + j\omega C}} = \sqrt{\frac{L}{C}}; \text{ for higher frequencies}$$

The losses associated with the transmission lines are:

- **Conductor losses:** due to resistance of conducting lines
- **Substrate losses:** dielectric medium between parallel conducting lines
- **Radiation losses:** accelerating changes at high frequencies resulting in electro-magnetic radiation.

Other losses such as attenuation losses, dissipation losses, and reflection losses are also part of conductor losses [5].

At high frequency, parallel lines are not used as radiation losses are more so various other transmission lines such as coaxial lines, microstrip, and waveguide are employed for the transmission of electro-magnetic energy. The main characteristics of the planar lines are compactness, lightweight, repeatability and conformity.

MIC and MMIC are generally employed upto 20 GHz and at higher frequencies waveguide structures or non-planar topologies are employed.

Circuits at radio frequencies can be configured as four-terminal network consisting of many resistors, capacitors, inductors, etc. and circuit analysis can be carried out by the measurement of input and output parameters [Annexure 1].

The measurement of four-terminal network employs various measurement techniques such as:

1. Z-parameter or open circuit impedance parameter
2. Y-parameter or short circuit parameter
3. H-parameter or hybrid parameters
4. S-parameters or hybrid open closed parameters

Z-parameters for four-terminals network can be represented as Z_{11}, Z_{12}, Z_{21}, Z_{22} indicate input, output, reverse and forward impedance; whereas Y-parameters such as Y_{11}, Y_{12}, Y_{21}, Y_{22} show short circuits parameters representing input, reverse, output and forward susceptance. H-parameters represent open and short-circuited parameters. S-parameters are employed at high frequencies where input and output waves are used for the circuit analysis [Annexure 2].

The input and output terminals are employed for finding out various parameters with the termination using the characteristic impedance. S-parameter for the two-port device are:

S_{11}: input scattering parameters

S_{22}: output scattering parameters

S_{12}: forward transmission coefficient

S_{21}: reverse transmission coefficient

Table 1.6: Mapping of distributed and lumped structure

Distributed transmission line	Equivalence (lumped element)	Representation
Short-circuited stub	Inductive ($\beta\ell<\lambda/4$)	$X=Z_0\tan(\beta\ell)$
Open-circuited stub	Capacitive ($\beta\ell<\lambda/4$)	$B=Y_0\tan(\beta\ell)$
Quarter-wave short circuited stub	Series L-C resonator	$L=\dfrac{\pi Z_0}{4\omega_0}; C=\dfrac{1}{\omega_0^2 L}$
Radial stub	Broadband short circuit	–
High impedance series line	Series inductance	$X=Z_0\tan(\beta\ell)$

1.4 Microwave Integrated Circuits (MIC)

MICs can be categorized into microwave printed circuit (MPCs) and hybrid MICs. The MPC technology is exclusively applied to a wide variety of microwave passive components including power dividers, filters, couplers, baluns, and printed antennas. The hybrid MIC technologies are commonly used for fabricating active, passive and integrated microwave circuits. Basic building blocks for MICs are planar transmission lines such as stripline, microstrip, proximity-coupled stripline, coplanar wave-guide and fin line. These lines are characterized by characteristic impedance, phase velocity, attenuation constant and power handling capability [7]. The two most common MIC technologies are thick and thin film. Thick film MICs are manufactured using various inks pressed through patterned silk screens and is employed at the lower frequency end. Thin film MICs are made via a sputter and etch process. The advantages of thin film circuit technologies are small size, light weight, excellent heat dissipation, repeatable performance and excellent RF characteristics at higher frequencies. As in any communication system, main building block is front-end consisting of filters, amplifiers, mixers, passive circuits which continuously demands shrinkage of size and better performance.

Various processes for the implementation of RF circuits are categorized as per below:

1. Printed circuit board (PCB)
2. Microwave integrated circuit (MIC)
3. Monolithic microwave integrated circuit (MMIC)

PCB are basically employed in electronics assembly and the main purpose is to provide mechanical stability and to provide multilevel inter-connection. This provides various interconnections between resistance, capacitance, inductance and other circuits. The main parameters for its selection are:

- Dielectric constant
- Attenuation
- Glass transition temperature
- Thermal expansion coefficient
- Cost effectiveness

Taking into considerations of the above parameters, the RF boards selected are mainly FR4 and PTFE where FR4 employed at low frequency (<3 GHz) and PTFE based material for low loss, high adhesive forces and moisture resistance connection.

MIC circuits can be segregated into two parts—hybrid integrated circuits and monolithic microwave integrated circuits [6]. The conductor and various transmission lines in hybrid integrated circuits are realized using photolithographic techniques. The active circuits in MIC such as transistor, diodes, etc. are soldered for the attachment. Hybrid MIC employs Alumina, Quartz and Teflon which are having low loss at higher frequencies (> 3GHz).

Monolithic microwave integrated circuits (MMIC) are realized on GaAs substrate having embedded active devices apart from passive elements. The circuits realized employing this process are resistance to thermal, shock and vibration. The circuits realized are compact due to same substrate realization apart from active and passive elements connection without any outside (parasitic) elements. MIC consist of various connection methodologies such as soldering, wire bonds, ribbon bonds in comparison to MMIC which are low loss, less radiation and compact but have limited output power due to associated thermal dissipation power output is limited. The circuits realized using above techniques are employed in wireless communication, aerospace and military applications.

Main fabrication steps in MIC fabrication are:

- Surface treatment of substrate
- Cleaning of substrate
- Metallization on both sides
- Photolithography and etching
- Circuits segregation
- Combination of elements
- Wire bonding
- Testing and characterization

Table 1.7: Comparative analysis of transmission lines

Description	Material	Outer Diameter	Power Capability (W)	Flexibility
Coaxial Lines	Teflon	0.14′	50	Semi-flexible
Coaxial Lines	Air	3.5′	12 K	Rigid
Waveguide (Al)	Copper	1′ × 0.5′	730 K	Rigid
Ridged waveguide	Aluminium	0.7′ × 0.4′	100 K	Rigid
Microstrip	Gold-plated alumina		50	Rigid

1.4.1 Microstrip Lines

Microstrip is the basic transmission line for the realization of planar circuits. Substrate selection is an important consideration for the realization of planar circuits on the microstrip. The conductor pattern having width ‘w’ is realized on the front side whereas back side is having complete ground plane. The main mode of propagation is the quasi-TEM and electric field ends at the ground plane also provide equi-potential surface. The main elements of microstrip are:

Dielectric substrate: It provides the stable base for the various structures such as transmission lines, resonant structures, planar antennas, etc.

Conducting layer: Top and bottom layers are having same metallization whereas top layer is etched as per the pattern and bottom layer is having continuous metal plane.

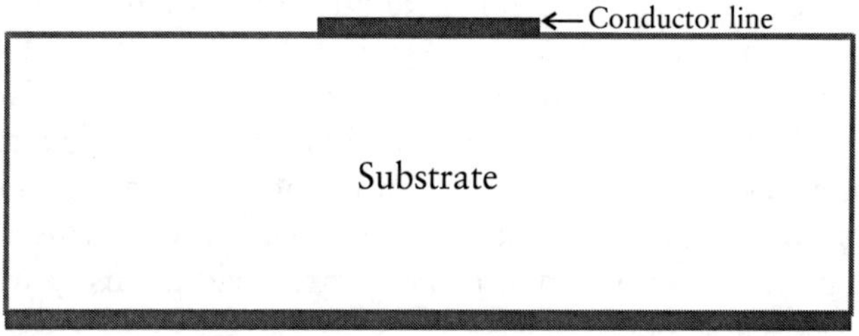

Fig. 1.1: Microstrip structure

Radiation loss is one of the contributors in signal attenuation and the design of the circuit need consideration so as to reduce or minimize the effect on the circuit performance. Main discontinuities such as open circuit and right angle bend are the major contributors [Annexure 3].

Substrate Thickness: At higher frequencies substrate thickness in microstrip to be lowered as surface wave losses become predominant in the dielectric. Above X-band, the surface wave losses associated with the 25 mils substrate are higher and recommended to have 10-15 mils substrate thickness to achieve the desired performance.

1.5 Permittivity Concept

According to Coloumb's law, the force between two charges are proportional to their amplitude and inversely proportional to the distance between them and also dependent on the nature of the medium. In insulating materials, the distance between conduction and valence band is high so the electrons transfer from valence band to conduction band is not possible. Microscopic local displacement is possible due to electric field and this insulating material is known as dielectric medium. Replacement of the medium in place of vacuum with another material, then the force between the charges are less compared to the vacuum.

The dielectric medium is represented by $\in$ which is known as dielectric constant.

ε_r = relative permittivity = permittivity of the medium(ε)/ permittivity in vacuum (ε_0)

The dielectric medium is represented by $\in$ which is known as dielectric constant. The relative permittivity or specific inductance capacity is varied in different mediums. Since each material is built using molecules and atoms which consist of electrons, protons and neutrons, so the effect or electric field displacing the orbital electrons which generate dipole field. This field resist the forced field resulting in generation of the polarization. In the case of linear, isotropic medium, the volume density of polarization is proportional to the electric field intensity which can be shown as below:

$$P = \varepsilon_0 X_e E$$

ε_0 = Free space wavelength

X_e= Electric susceptibility

D = Electric Flux density

This relation can be represented as

$$D = \varepsilon_0 E + P = \varepsilon E = \varepsilon_0 \varepsilon_r E$$

$$\varepsilon = \varepsilon_0(1 + X_e) = \varepsilon'(1\text{-}j\tan\delta)$$

The above equation shows the relations between permittivity and electric nature. The imaginary part for the lossless material is taken as zero. The value of tanδ is kept constant (loss tangent) and it represents angle between phase and total current.

Table 1.8: Permittivity value and tanδ of some material

Material	Permittivity	Loss tangent (tanδ)
Alumina	9-11	0.0006
Glass (ceramic)	6.0	0.0050
Quartz	4.0	0.0001
Teflon	2.1	0.0003

Another expression for the representation of the relative permittivity is:

$$\varepsilon = \varepsilon' - j\varepsilon'' = \varepsilon'(1 - j\tan\delta)$$

$$\varepsilon_r = \varepsilon_\infty + \left[\frac{\Delta\varepsilon_i}{\left(1 + j\frac{f}{f_c}\right)}\right] - j\frac{\sigma}{\omega\varepsilon_0}$$

$\Delta\varepsilon_i$ = ith..Debye..function
f_c = relaxation..frequency
σ = conduct...at...zero...frequency

The relative permittivity can be expressed as

$$\varepsilon_r \frac{\varepsilon' - j\varepsilon''}{\varepsilon_0} = \varepsilon_r' - \varepsilon_r'' = \varepsilon_r'(1 - j\tan\delta)$$

In case of semiconductor, charges are free and bounded which contributes the current by applying electric field. The concept of permittivity can be deduced from the Maxwell's equation [8]

$$\nabla XH = J = J_D + \frac{\partial D}{\partial t} = j\omega D + J_C = \sigma E + j\omega\varepsilon_0(1 + \chi_e)E \quad \text{... (1.11)}$$

$$= [\sigma + j\omega(\varepsilon' - j\varepsilon'')]E$$

$$J = j\omega\left(\varepsilon' - j\varepsilon'' - j\frac{\sigma}{\omega}\right)E = j\omega\varepsilon^* E$$

$$\varepsilon^* = \left(\varepsilon' - j\varepsilon'' - j\frac{\sigma}{\omega}\right); \text{ complex.dieletric.constan } t \quad \text{... (1.12)}$$

for $\sigma = 0$

$$D = \varepsilon E = \left(\varepsilon + \varepsilon_1\frac{\partial}{\delta t} + \varepsilon_2\frac{\partial^2}{\delta t^2} + \ldots\right)E \Rightarrow D = (\varepsilon + j\omega\varepsilon_1 - \omega^2\varepsilon_2 + \ldots)E$$

$$\Rightarrow \frac{D}{E} = \overline{\varepsilon}(\omega)$$

where $\overline{\varepsilon}(\omega) = \varepsilon' - j\varepsilon'' = |\varepsilon|e^{-j\delta} = \varepsilon_0(1 + \chi_e)$

$$\tan\delta = \frac{\omega\varepsilon'' + \sigma}{\omega\varepsilon'} \approx \frac{\varepsilon''}{\varepsilon'} \quad \text{... (1.13)}$$

$$Q = \frac{1}{\tan\delta} \approx \frac{\omega\varepsilon'}{\omega\varepsilon'' + \sigma}$$

The imaginary part associated with the permittivity accounts for losses in the medium due to damping of the vibrating dipole moment [8]. Dielectric loss due to damping effect can be further correlated with electric susceptibility and higher susceptibility renders more heating. In the case of semiconductor $\sigma = q\mu n$ so for frequency '*f*' using eq. (1.13)

$$\tan\delta = \frac{(\omega\varepsilon'''' + q\mu n)}{\omega\varepsilon'} Q = \frac{G}{\omega C} = \frac{\text{conductance}}{\text{capacitance}}$$

The generalized conductivity can be represented as

$$\sigma = nq\mu_n + pq\mu_p \approx nq\mu$$

In case of microstrip line mostly for narrow lines [9], the relative permittivity can be represented as

$$\varepsilon_r = \frac{C_d}{C_a} = \frac{c}{v_p^2} = \frac{\text{velocity of light}}{\text{phase velocity}}$$

where c is the velocity of light, C_a is air capacitance, C_d is the dielectric capacitance.

Variation of tanδ with frequency is more prominent compared to ε_r at microwave ranges. Lossy dielectric such as silicon having higher dielectric susceptibility leads to more polarization rather than increasing current density with higher electric field. The losses in the dielectric can be represented by generic term of power factor which represents power factor of condenser containing the dielectric

$$\text{Power factor} = \cos(90° - \delta) = \varepsilon_r = \frac{\varepsilon''}{\sqrt{(\varepsilon')^2 + (\varepsilon'')^2}};$$

It can be equated to tanδ for $\varepsilon'' << \varepsilon'$

Permittivity of any material can't be lesser that ε_0 and classification of perfect dielectric, good dielectric and lossy dielectric [9] can be represented in RF ranges as shown in Table 1.9.

Table 1.9: Classification of dielectric material in terms of permittivity

Dielectric	Permittivity behaviour
Perfect	$\varepsilon'' = 0$
Good	ε' remains constant, ε'' considerably small and increase with frequency
Lossy	Both ε' and ε'' varies with frequency

Generally in case of pure semiconductor, dissipation factor is taken as ~1 which indicated conduction current density equals to the displacement current density ($J_c = J_d$; $\sigma = \omega\varepsilon$).

1.6 Fabrication Aspects

Fabrication of MICs can be divided into two parts: soft substrates assembly and realization of MIC on ceramic or semiconductor substrate mainly on silicon. The assembly over soft substrate are comparatively easy considering planar transmission media such as microstrip or CPW. The basic substrate is laminated with conductor of appropriate thickness using adhesive material. The microstrip components are then fabricated using photolithography. Final circuits are then assembled using soldering technique as employed in PCB assembly.

1.6.1 Substrate Selection

Substrates employed in MIC should possess various properties such as: minimal loss at microwave frequencies, better adhesion with conductors, smooth polished surfaces, avoidance of voids free and no deformation during various processing steps. The top metallization layer is having the Cr-Cu-Au: 200-300Å, Cu:2 um, Au-3-5 um to cater for RF applications. The metallization layers play significant role where hot plate operations are carried out which can be altered to either TiW, Au, Cu, Ni, Au or Cr, Cu, Ni, Au. The main criteria of an ideal substrate material is: high dielectric constant, low dissipation factor (tanδ), constant dielectric value over temperature ranges, high purity, constant thickness, high surface smoothness, dielectric strength and thermal conductivity [3].

1.6.2 Fabrication Aspects

The major steps of MIC fabrication using thin-film technology are as appended below:

- Photomask fabrication
- Substrate processing
- Substrate metallization
- Photolithography
- Assembly and packaging

The above steps are mostly employed for the fabrication of MIC.

1.6.2.1 Photomask Fabrication: Mask are made on either plastic films or glass plastic films or glass plates having positive/ negative MIC Pattern. The manual method uses co-ordinatograph and reduction camera combination for this purpose. Co-ordinatograph is used to produce enlarged pattern (10 X or else) which is then reduced to the actual (1X) size using reduction camera (stepper). This method is capable of making masks with an overall accuracy in microns.

Substrate processing refers to cutting the substrate to required size and drilling holes in it and possible to perform batch processing of MIC substrates. In such a case dicing has to be performed after circuit is patterned on to the substrate otherwise dicing is performed before photolithography.

Sawing and laser cutting normally achieve very accurate edges. Sawing is slow and some substrate is lost in the process. Laser is a very fast process and capable of automation diamond scratching is also very fast accurate and cuts the substrate without material loss. In drilling laser beam is fastest gives very good quality of edges and capable of drilling holes of non-round shape. Diamond drilling is slow but produces good quality of round holes. Ultra sonic also produce acceptable quality of hole and can be used to non-round hole. Band blasting result in funnel-shaped holes.

1.6.2.2 Substrate Metallisation is the most important process as the MIC electrical performance reliability singularly depends on metallisation. This process is accepted to produce well adhering metal thin films on the substrate with various properties as close as possible to the bulk material. Sputtering and vacuum evaporation techniques are generally used for substrate. Metallisation followed by chemical plating for increasing thickness and out of these two techniques sputtering is accepted to provide metallised material having properties more closer to the bulk material. Techniques such as magnetron sputtering and ion cluster deposition are also reported in literature which is reported to give excellent film quality. Both of these techniques are capable of achieving fast deposition rates. It is therefore possible to all together avoid wet chemical deposition and get excellent quality of metal thin films avoiding thin film stress.

Thin film stress: Stress can be present as an intrinsic stress in the substrate due to process parameters or can be introduced by external means mainly due to thermal expansion of the films. The mean stress in thin films is due to differences in the coefficient of expansion of the film and substrate generated stress due to cool-down phenomena from the deposition temperature. CTE mismatch of the thin film results in an extension or contraction and excessive compressive or tensile stress results in buckling, cracking, sticking and splintering problems. The stress generation in the layers can be related with the bow or curvature phenomena. Curling of thin film layers, misalignment of device features, resonant frequency shifts are directly linked to stress. Stress is predominantly affected by the thickness and deposition conditions. Stress is categorized as tensile or compressive stress and can cause fracture, blistering, buckling or curling in the device due to stress in these layers. Stress gradient originates from the variation of stress to the growth surface due to grain size variation with thickness. Higher thickness can lead to lower stress but it cause subsequent wafer distortion such as bow and cracking prohibiting subsequent wafer processing involving photolithography steps [10]. Another parameter is temperature as thermal stresses are the highest contributors to the overall stress value. Annealing reduces the stresses caused by mismatches expansion and by non-uniform nucleation of the film layer.

1.6.2.3 Photolithography process involves transfer of photo mask pattern on to the substrate using photosensitive film called photo-resist. The underside metal is then etched out in case of shadow printing and extra material is deposited in case of selective e plating technique. The selective e plating process results in a very good edge definition compared to shadow printing. Main steps in lithography are: Priming, soft baking, hard baking, exposure, developing. Priming is carried out to promote adhesion, soft baking for the removal of solvents from the PR and hard baking for the improved adhesion [11].

1.6.2.4 Assembly and packaging of MIC involves attaching active and passive components of MIC substrate, mounting of

MIC substrate on the carrier plate, assembly of carrier plates in the box/package, interconnection of MIC substrate, attachment of connectors.

1.6.2.5 MIC fabrication on Silicon: After subjecting to standard thin film substrate cleaning cycles, silicon substrates are sputtered with thin layer of TiW (200-300Å) followed by 8000Å of gold film on both sides of substrates. This combination of seed layers is electroplated with gold to the required thickness of 4.5 μ ± 3% and circuits are patterned using standard optical lithography and subtractive etching process. The patterned substrate has been attached to gold metallized Kovar carrier plate using silver-based conductive epoxy. The carrier plate is mounted in test jig and RF connectors are connected by gold ribbon using parallel gap welding.

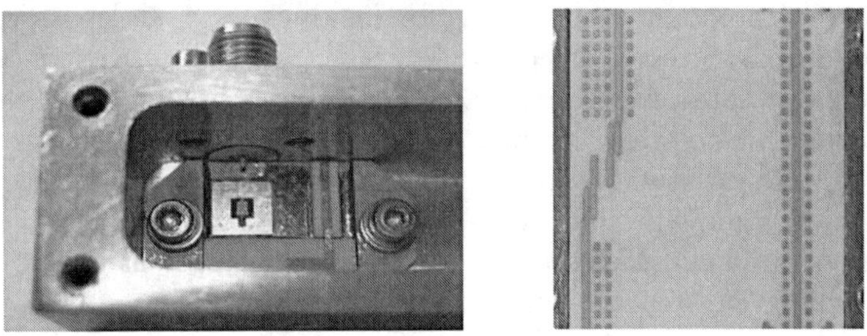

Fig. 1.2: Complete MIC assembly

References

1. D.M. Pozar, *Microwave Engineering*, John Wiley & Sons.
2. G.J. Wheeler, *Introduction to Microwaves,* Prentice Hall, 1978.
3. S.Y. Liao, *Microwave Devices and Circuits*, Prentice Hall.
4. T.C. Edwards *Foundations for Microstrip Circuit Design*, John Wiley & Sons.
5. K.C. Gupta and Garg, Chaddha, *CAD for Microstrip Circuits*, Artech House, 1993.
6. Singh, Kamaljeet, *Principle of Radio Frequency and its Application in RF-MEMS,* Hindi Publications, Prabhat Prakashan, 2018.
7. Bahl, Inder, *Lumped Elements for RF/Microwave Circuits'* Artech House, London 2003.
8. Singh, Kamaljeet, *Interdigital Based Capacitor*, Atlantic Publishers and Distributors (P) Ltd., 2020.

9. Matthew, N.O. Sadiku, *Elements of Electromagnetics*, Oxford University Press, 2001.

10. Sedky, Sherif, *Post-processing Techniques for Integrated MEMS*, Artech House 2005.

11. Fooks, E.H. and R.A. Zakarevicius, *Microwave Engineering using Microstrip Circuits*, Prentice Hall, 1990.

Process contamination controls are important for assembly and packaging of the microwave circuits having active devices which are prone to ESD, temperature and handling. Contaminants such as dust, rust or oxidation has detrimental effect on the performance. The ceramic substrate such as alumina should possess chemical purity, tensile strength, better surface finish, precise density and dielectric properties. The handling of ceramic is also pivotal as it is brittle and can chip or crack.

Chapter 2

Multipliers Basics

2.1 Introduction

Multipliers can be designed with the diode or transistor keeping into considerations various parameters and applications. The important parameters are cost effectiveness, planarity, multiplication factor, ease of implementation and system level requirements. Solid state-based multipliers are compact, lightweight, reliable and inexpensive compared to vacuum tube technology. The main characterization parameter is the harmonic content versus input frequency and power level. RF waves constituted of series of fundamental and harmonic frequency terms which add together to form complex waves and can be analyzed in the time domain or frequency domain. In non-linear circuits, generation of frequencies are different from the input frequencies.

$$v_{out} = a_0 + a_1\, v_{in} + a_2 v_{in}^2 + a_3 v_{in}^3 + a_4 v_{in}^4 + \ldots$$

The first two terms are linear and other terms are non-linear. a_0 is DC term whereas a_1, a_2, a_3 are constant and v_{in} is the input voltage.

In case of $v_{in} = v \sin \omega t$ (taking only first two terms)

$$V_{out} = a_0 + a_1 v \sin \omega t + a_2 (v \sin \omega t)^2$$

$$= a_0 + a_1 V \sin \omega t - a_2 v^2 (\cos 2\omega t)^2/2 + a_2 v^2/2$$

Additionally, DC term, fundamental frequency and harmonic term 2ω are generated in the non-linear circuit. In case of two frequencies ω_1 and ω_2, the input voltage v_{in} can be written as

$$v_{in}(t) = v \sin \omega_1 t + v \sin \omega_2 t$$

$$v_{out}(t) = a_0 + a_1[v \sin \omega_1 t + v \sin \omega_2 t] + a_2 [v \sin \omega_1 t + v \sin \omega_2 t]^2$$
$$= a_0 + a_1 v \sin \omega_1 t + a_1 v \sin \omega_2 t + a_2 V^2 (1\text{-}\cos 2 \omega_1 t)^2/2 + a_2 v^2 (1 - \cos 2 \omega_2 t)^2/2 + a_2 v^2(\cos \omega_1 t - \cos \omega_2 t) - a_2 v^2 (\cos \omega_1 t + \cos \omega_2 t) + \ldots$$

So various frequency components which are generated such as DC term (a_0), fundamental frequency terms (ω_1 and ω_2), harmonic terms ($2\omega_1$ and ω_2) along with other harmonics and mixing terms. Taking $v_{in}(t) = x \sin \omega_1 t + y \sin \omega_2 t$ than the harmonic component amplitude can be xa_1, ya_1, $a_2 xy/2$, $a_2 xy/2$, $a_2 x^2/2$, $a_2 y^2/2$, $a_3 x^3/4$, $a_3 y^3/4$ for the harmonic component ω_1, ω_2, $\omega_1 - \omega_2$, $\omega_1 + \omega_2$, $2\omega_1$, $2\omega_2$, $3\omega_1$, $3\omega_2$.

The various multipliers approaches stem out from the fact that any non-linear device and can be represented as:

$$V_0 = \sum_n a_n V_{in}^n = a_1 V_{in} + a_2 V_{in}^2 + a_3 V_{in}^3 + a_4 V_{in}^4 + \ldots\ldots$$

In case of doubler first two terms are taken into consideration and for tripler first three terms are considered [1-2]. Frequency multiplier can be built from a non-linear electronic component which generates a series of harmonics, followed by a band pass filtering to pass the desired harmonics to the output while blocking other components. The non-linear circuit should maximize the coefficient for the desired harmonic and minimize the others for maximizing the efficiency. Generally, an even function is chosen to generate even harmonics or an odd function for odd harmonics [1].

Schottky barrier varactor and step recovery diode are employed for multipliers operations. Schottky barrier is employed for RF to the sub-millimeter ranges, varactor (p-n/Schottky junction) upto several GHz whereas SRD typically upto 18 GHz. FET family such as metal epitaxial semi-conductor field effect transistor (MESFET), high electro mobility FET (HEMT), metal-oxide FET are employed for mixers apart from usage in amplifiers, oscillator, switches, attenuators, etc. Various multiplier topologies employing varactor, SRD, Schottky diode and FET (Table 2.1) can be constructed at high frequencies [2].

Diodes	Applications	Characteristics
Varactor	Modulators Harmonic generators Parametric Amplifier Upconverters Tuning elements	Variable junction capacitance $C_j = \dfrac{C_{j_0}}{\left(1-\dfrac{V}{\phi}\right)^{\gamma}}$
Step recovery (SRD)	Harmonic generator	Short circuit in forward bias Capacitance in reverse bias
PIN	Switch Pulse modulator Phase shifter	Attenuating Levelling Amplitude modulation

In cases of varactor diode, the phase changes of the modulating microwaves versus bias voltage of the diodes are non-linear because of the diode capacitance. Non-linear phase change may cause transient jittering and bit error rate degradation. Schottky barrier diodes are considered to be suitable for obtaining excellent modulated waves in a low power region.

Multiplier operation depends on the non-linear properties of Schottky diodes having capacitive and resistive non-linearity. A non-linear resistor or capacitor driven by source can generate series of harmonic frequencies which is the basic form of harmonic multiplier. This behaviour can be employed for frequency multiplier applications. Change in the non-linear capacitor can be represented as

$$Q(V) = \sqrt{a(\Phi_{bi} - V)}$$

where a is constant, ϕ_{bi}-built in potential

The function can be extended as

$$Q(V) = \frac{dQ}{dt} = [a_1 + 2aV + 3a_2V^2 + \dots]\frac{dV}{dt}$$

Manley-Rowe equations show that ideal frequency multiplier at 100% efficiency converting input power to higher frequency output power by properly terminating all other frequencies. Non-linear resistors used as frequency multiplier can be theoretically 25% for doubler and 11% efficient for tripler applications.

The capacitor displacement current is shown as

$$Q = CV$$

$$I = \frac{dQ}{dt} = \frac{dC(V)V(t)}{dt}$$

$$V = V_0 \sin \omega t$$

$$I(t) = C(V)V_0\omega\cos\omega t + V_0 \sin\omega t\frac{dC(V)}{dt}$$

$$= V_0\omega C(V)\cos\omega t + V(t)\frac{dC(V)}{dt}$$

This assumption is based on assumption that displacement current flowing through the capacitor must be equal to the conduction current flowing through the un-depleted resistive region.

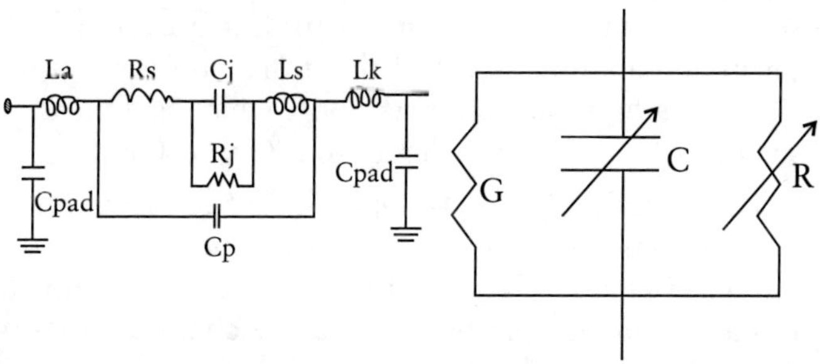

Fig. 2.2: Generic diode model

The diode model consists of a series Rs and non-linear resistance (PIN diode) and non-linear capacitance (varactor diode). The value of Rs and Cj are dependent on the doping density, series resistance Rs is mainly contributed by the un-depleted epi-layer. Circuits such as mixers, multipliers are performance limited due to circuitry rather than diode performance [4].

2.3 Diode Parameters

Carrier Life Time: An important parameter of diode is the carrier lifetime, which is useful for defining the low frequency limit $f_o = \pi/2$ for the linear performance of the diode. For RF signal frequencies below f_o, the PIN diode rectifies the signal much like an ordinary PN junction diode and results in considerable output distortion. At frequencies above f_o, less rectification occurs with increasing frequency, allowing the pin diode to appear more linear, approaching a pure resistor. For applications requiring good linearity and low distortion, the minimum signal frequency should be ten times f_o, i.e. $f_{min} = 10f_o$. This restriction is not important in switching applications where the diode is normally either completely ON or OFF. In those states, since most of the power is either reflected or transmitted, the effect of the RF current on the total charge is small and the distortion is not a problem.

Reverse Recovery Time: It is a measure of the switching time and is dependent on the forward and reverse bias applied. With forward bias current, charge is stored in the I layer and when a reverse pulse is applied, reverse current flow for a short period of time, known as the delay time (t_d). The current begins to decrease when sufficient number of carriers are removed. The time required for the reverse current to decrease from 90% to 10% of its initial value is called the 'transition time' (t_r). The sum t_d+t_r is the "reverse recovery time", which is a measure of the time taken to switch the diode from ON to OFF.

Reverse Break Down Voltage: It is the recommended maximum signal level for the safe operation of the diode. Operation at signal levels above the reverse breakdown voltage may result in the degradation of the diode characteristics or in permanent damage to the diode.

Insertion Loss: The loss of the signal attributed to the diode when the switch is ON (transmission state) is called "insertion loss". For low insertion loss, low resistance the series switch and low capacitance in the shunt switch are needed.

Isolation: Isolation is a measure of RF leakage between the input and the output, when the switch is OFF. For high isolation (low transmission), low capacitance is required in a series switch especially at a higher frequency.

Cut-Off Frequency: Mixer can also be used for the generation of the various harmonics with the bottleneck of employing two sources. Basic non-linear element in mixer is responsible for the generation of various frequency components. The mixer construction can be made by the diode or transistor-based topology. The diode-based topologies due to ease of implementation; repeatability and higher multiplication factor are preferred. In this section, diode-based topology is discussed employing Schottky diode.

$$f_c = \left(\frac{1}{2\pi R_s C_{jv}} \right)$$

The cut-off frequency for silicon diodes upto 250 GHz and for GaAs diodes upto 900 GHz but the operation is generally limited to $f_c/10$. The frequency of operation beyond $f_c/10$ increases the Rs as well as noise.

The governing relation for the non-linear relation between current and voltage can be expressed as:

$$I = I_s \left(e^{\frac{q(V-IR)}{nkT}} - 1 \right)$$

The R_s represents bulk resistance of the semi-conductor and contact resistance. N is the ideality factor and value taken is ~ 1.0.

Switching Speed: Reverse recovery time is a measure of the switching time.

Power Handling Capability: The RF power that can be safely handled by the diode switch is limited by two factors.

1. The breakdown voltage of the diode.
2. Thermal considerations.

Thermal considerations involve maximum junction temperature and thermal resistance of the diode and packaging. Other factors affecting power handling capability are ambient temperature, frequency, attenuation level (which is related to diode resistance), pulse width and duty cycle [4].

Maximum power a diode can dissipate is given as

$$P_{max} = \frac{T_{j(max)} - T_A}{\theta_{jc} + \theta_j}$$

$T_{j,\,max}$-maximum operating diode junction temperature

T_A-ambient temperature

θ_{jc}-Thermal resistance-junction case

θ_{jA}-Thermal resistance-case to ambient

2.4 Varactor-based Multipliers

Varactor is a non-linear reactive device exhibiting voltage dependent capacitance used for harmonic generated parametric amplification, mixing, detection and voltage variable tuning applications [Fig. 2.3]. The most prominent applications of varactors is the harmonic generation at millimetric and sub-millimetric wave frequencies.

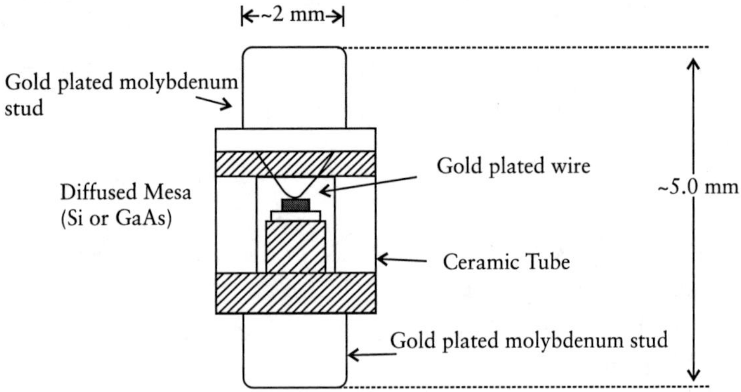

Fig. 2.3: Varactor construction

The varactor model consists of a series R_s and non-linear reactances (C_j and R_j) as shown in Fig. 2.4.

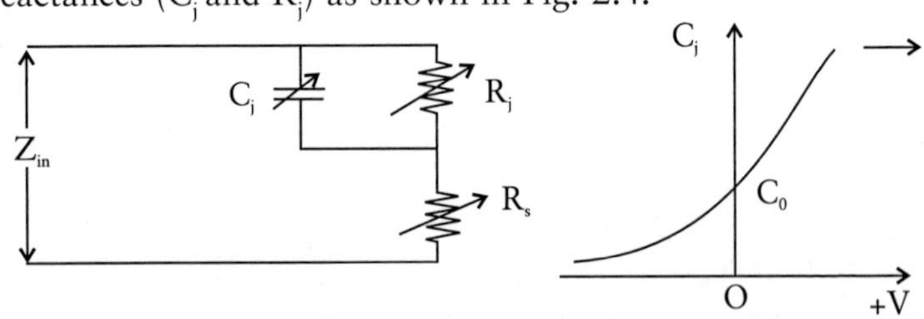

Fig. 2.4: Varactor equivalent circuit   Capacitance variation

Varactor is a non-linear reactive device exhibiting voltage dependent capacitance used for harmonic generated parametric amplification, mixing, detection, pulse generation, pulse shaping and voltage-variable tuning. The most prominent applications of varactor is the harmonic generation at millimetre and sub-millimeteric wave frequencies. The varactor model consists of series resistance R_s and non-linear differential resistances.

A varactor diode multiplier also relies on capacitive non-linearity which is the function of the reverse voltage is the diode. It provides relatively low conversion loss and excellent phase noise characteristics [5]. This aspect along with high efficiency makes it ideal for multiplier applications. However, the varactor multiplier is narrowband, highly sensitive to circuit parameter variations and having low stability. Also at the input frequency the associated susceptance to be high enough to sustain reactive current. As small variation in circuit parameters exhibit negative resistance and can become unstable resulting in variable output power. Due to above-mentioned reasons, the varactor multiplier is rarely used in modern synthesizer designs [6, 7].

2.5 PIN Diode

The most important feature of the pin diode is its property of being almost pure resistor at RF frequencies, whose resistance value can be varied from 10000 to less than 1 Ω by the current flowing through it along with linearity, low distortion and a low current drive. PIN diode appears as a pure resistance at RF whose resistance can be varied from 1-10 kW by controlling current and basically employed in attenuator and switching applications.

The performance of the pin diode circuit is directly related to the basic characteristics of the diode. The performance of the pin diode switch can be simply approximated by treating the pin diode essentially as a resistor in the forward-biased state and a capacitor in the reverse-biased state. Switching performance can then be analyzed by calculating the insertion loss, isolation, switching speed and power handling capabilities.

2.5.1 PIN Diode and its Equivalent Circuit

PIN diode behaves as an ordinary diode upto 100 MHz and at higher frequencies it acts like a variable resistance. The behaviour of the pin diode at microwave frequencies however is entirely different from that at normal frequencies. It can be used as a switch (resistance variation) when operated between fixed forward- and reverse-biased states for switching microwave signals. In case of the forward DC-biased application to the pin diode free charges from the P and N regions flood the I region, thus converting into conductive medium the diode behaves like a short circuit and any microwave signal super imposed on it passes without any rectification.

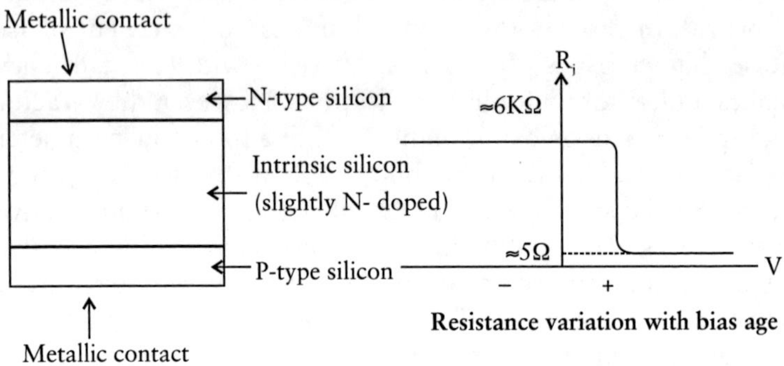

Fig. 2.5: P-I-N Diode

In case of reverse bias application, the I region gets completely depleted of charge carriers and the diode behaves virtually like an open circuit having a large capacitive reactance.

In case of microwave signal super imposed on the DC voltage, the time period of the positive half cycle would be too short when compared with the lifetime of the charge carriers. This forbids any conduction from taking place in the I region. Thus, for the microwave signals, the pin diode act as linear device for which impedance is governed by the slope of the DC characteristics at the operating point.

An equivalent circuit can represent a packaged pin diode where L_p and C_p are the stray inductance and capacitance. The typically values of L_p and C_p are in the ranger 0.1 to 2 nH and 0.1 to 0.5 pF respectively.

The parameters R_j, C_j and R_b represent the junction resistance, the junction capacitance and the bulk semiconductor resistance of the diode. The equivalent circuit of a pin diode under forward bias can be represented by a small resistance of the P and N layers and the contact resistance. The value of R_f is typically in the range 0.2 to 5 under reverse bias, the pin diode can be represented by a capacitance C_j in series with a small resistance R_r. The typical values of C_j and R_r are in the range of 0.001 to 2 pF and 0.2 to 5 W respectively.

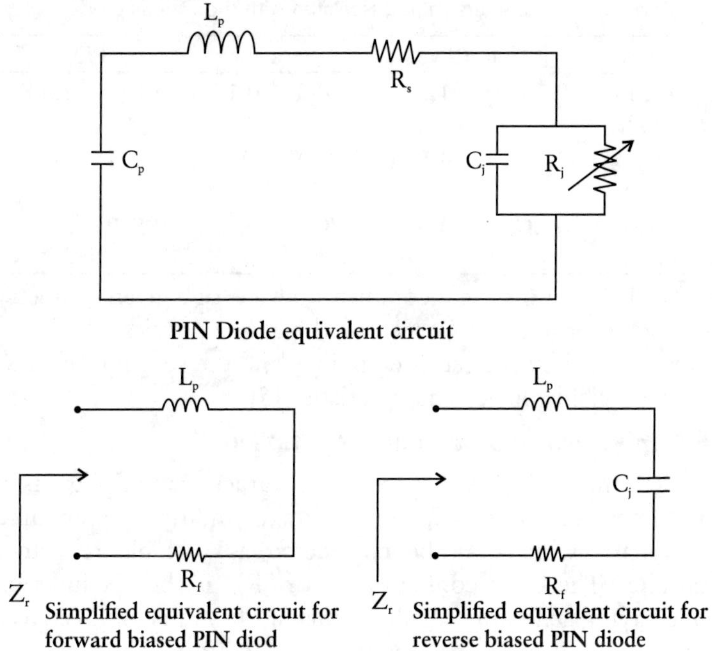

Fig. 2.6: Equivalent model of the P-I-N diode

PIN application is in the switching, pulse modulation, attenuation, pulse shifting of an RF signal. It appears as pure resistance at RF for which resistance varies from 1-10 kΩ by the current control [8].

Choice of the diode: Schottky diodes Vs PIN diode: A beam lead Schottky diode with excellent RF characteristics is need to; achieve strong reflections in forward and reverse states. Most commercial Schottky diodes have a dynamic resistance of 15 to 20 Ω giving reflection coefficient of –2.7 and –3.7 dB respectively. On the other hand, a beam lead PIN diode will have a resistance as low as 5 Ω (5 ohm) giving a reflection coefficient of –0.9 dB. The package capacitance of both diodes lies between 0.12 to 0.15 pF. The value of the reflection coefficient indicates that the PIN diode is a better choice for the modulator this because a higher value of the reflection coefficient results in a greater reflected power so PIN diode is chosen for the QPSK modulator for having better performance.

Table 2.3: Parameters value associated with the various diodes

Values	Schottky		Varactor		PIN	
C_j (junction capacitance)	0.36pF	0.12-0.15pF	0.36pF	0.15-0.7pF	0.1 pF	20pF
R_j (junction resistance)	270Ω	5-17kΩ	10MW	10MΩ	20kΩ	0.8Ω
R_s (series resistance)	4-6Ω	4-10Ω	0.7Ω	0.4Ω	0.5-2Ω	0.5-2Ω

The behaviour of semiconductor device for a given geometry and doping profile is governed by the device physics which can be represented by the set of equations and the solution of these equations yields device characteristics [5].

2.6 Step Recovery Diode (SRD) Multipliers

SRD multipliers differ from the varactor multiplier as it can be considered as switching reactance multiplier compared to varactor which can be considered as variable reactance multiplier. The basic configuration of SRD is the PN junction diode having generation of short pulses. They are employed as a comb generator so can be used for the generation of higher order harmonics (> 20). There efficiencies are higher as they are having little capacitance change with the frequency. Transition time is the key parameters associated with the SRD which indicates transition from low impedance state to high impedance state resulting in narrow pulse of periodic voltage.

A step recovery diode (SRD) has long been employed for the realization of high order frequency multiplier. The SRD utilizes capacitive non-linearity to generate a train of very short pulses that is rich in high order harmonics. The circuit is known as comb generator and is often used in multiloop synthesizers to generate high frequency LO signal with a course step size. A fixed frequency filter are switched filter bank and can be placed at the SRD multiplier output to choose a desired harmonic, as it is often used in direct analog synthesizer designs [9].

SRD operates in two states where in the forward bias it looks large capacitor having low impedance and under reverse bias it looks like small capacitor having high impedance. It is three-layer device and intrinsic layer charge can be injected and stored. Basic operation is based on the charge injection in the intrinsic layer and removal of the charges resulting in sudden increase of the diode resistance resulting in the voltage spike. This narrow voltage pulse generated by sudden change of current is equivalent to the comb generation (output frequencies) which is multiple of the input frequency. The equivalence of the SRD can be short circuit when forward biased storing all forward current whereas in reverse bias condition it has constant capacitance [10].

SRD is also a diode with a PN junction. The reverse recovery time t_{rr} when switching from ON to OFF is short, and therefore, the transfer time in which the characteristic is rapidly turned off is remarkably short. Application of the sine wave is applied to the step diode, since the tt (transfer time) is short, the output waveform is suddenly pinched off, so that many high-frequency harmonics are generated. Its structural characteristics are: there is a steep impurity distribution area at the boundary of the PN junction, thereby forming a "self-service electric field". Since the PN junction is under forwarding bias, it conducts with minority carriers and has a charge storage effect near the PN junction, so that its reverse current needs to undergo a "storage time" before it can be reduced to a minimum (reverse saturation current value). SRD beside employing for higher order multiplication in aerospace systems is also suitable in applications such as drone detection system monitoring all three of the bands, 433 MHz SRD and 2.4 or 5.8 GHz ISM, while skipping over the vast spectrum in between.

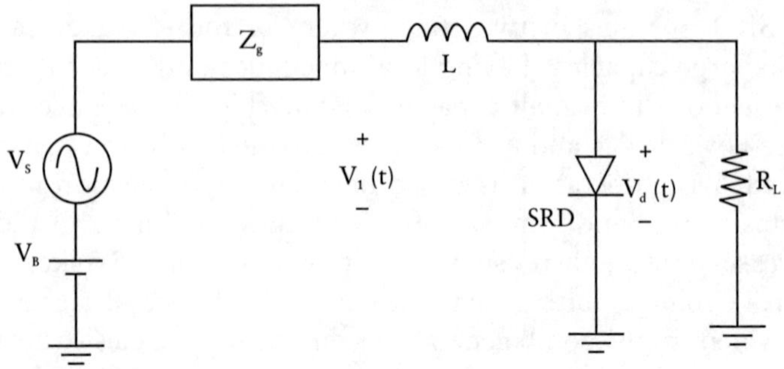

(a) SRD Equivalent

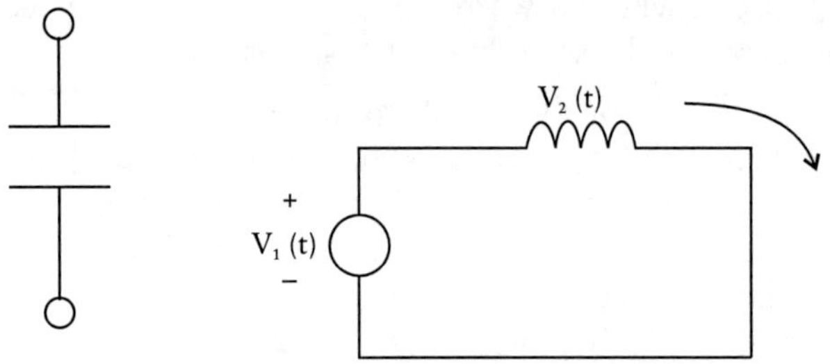

(b) Forward bias (C~∞)

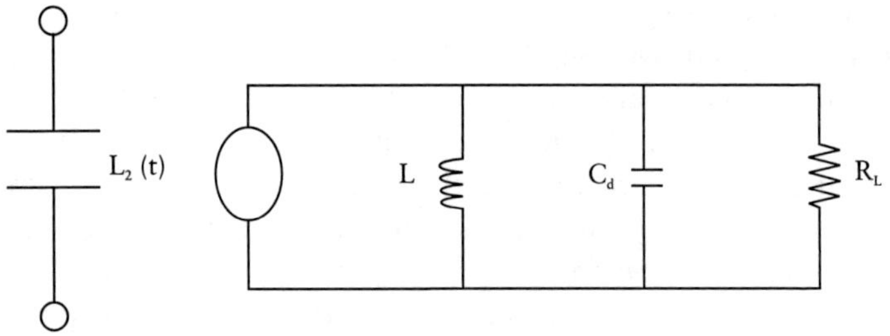

(c) Reverse bias(C~finite)

Fig. 2.7: SRD equivalence

An ideal SRD is having high breakdown voltage (25-110 V), small reverse bias capacitance (0.8-6.5 pF), small series resistance (0.12-0.8 Ω), high minority carrier life time (τ > 100 ns), small transit time (< 200 ps) and small thermal resistance (20-50° C/W). The Ohmic and Schottky contact on semiconductor [4] can be carried out employing various metallization scheme as shown in Table 2.4. Silicon and GaAs is normally employed and silicon is preferred due to power handling capability, process stabilization and high resistivity in the intrinsic region [4].

Table 2.4: Metallization scheme

Contact	GaAs	GaN
Ohmic	Au/Ge/Ni, In, Sn, Ag	Ti/Al/Ni
Schottky	Ti/Pt/Au	Ni/Au/Ti

Multiplier (and comb generator) circuits are reactively terminated at the higher-order harmonics at the output port resulting in the power partially reflected and recombined to produce stronger harmonics.

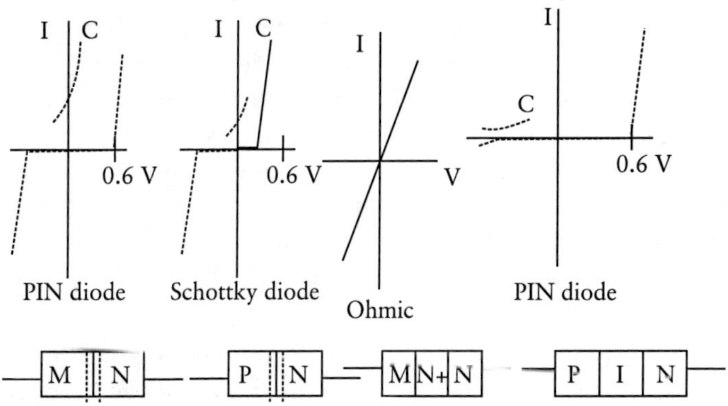

Fig. 2.8: Various diodes and their characteristics

2.7 Transistor Multipliers

The signal generation in the mm wave bands can be realized with multiplication from a lower frequency to have superior noise performance in high quality oscillators' realization. Active multipliers has the inherent advantage over diode multipliers by having lower conversion loss, inherent isolation from input to output, temperature stability, no need of idler circuits and much lower drive power for generating significant output power level [11]. High electron mobility transistor (HEMT) has been chosen generally for having better phase noise characteristics at higher frequencies. Active multipliers have been configured as frequency doublers mostly and single stage FET Tripler at higher frequency has been reported very few. Active multiplier with an order greater than two is difficult to realize at higher frequencies due to parasitic feedback, dispersion and efficient square wave generation. Also the terminating impedances at the fundamental frequency and harmonics become interrelated. This is either due to unavailability of model or circuit complexity which makes mismatch between the simulated and practical results.

The salient features of the multipliers are Class A operation with 0 dBm (nominal) input power with higher efficiency.

Transistor multipliers use both bipolar and FET devices [12]. Device is biased in such a manner that it conducts during a portion of the input sine wave cycle. Similar to diode multipliers the signal waveform is distorted and harmonics are created. The transistor DC bias is adjusted to accentuate a selected harmonic. A balanced design is also possible as which utilizes two devices that are fed in anti-phase and conduct current during a positive and negative half cycle respectively. The circuit delivers pulses per period of the incoming signal therefore generating a strong second harmonic. All odd harmonic products should be generally suppressed. The level of separation depends on the mismatch between individual transistor characteristics.

The main advantage of a transistor multiplier is its high efficiency. Non-linear behaviour and signal amplification functions of a transistor are combined to achieve frequency multiplication with minimal conversion loss or even conversion gain. This minimises the number of additional components (such as amplifiers) required to compensate for signal loss. The main disadvantage is higher noise which can restrict its application in multiplying a low noise reference signal.

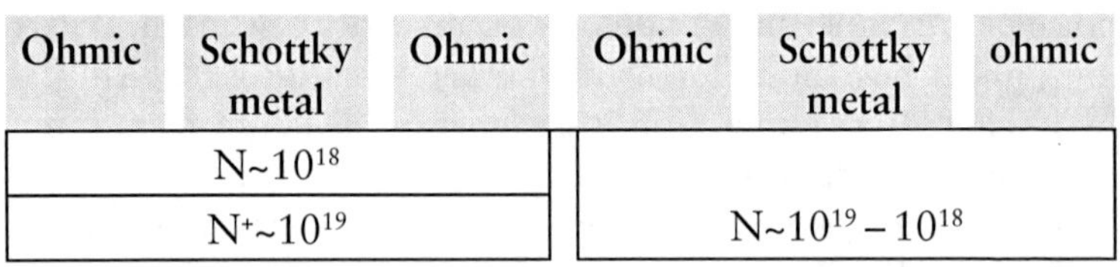

(a) (b)

Fig. 2.9: Construction schematic of (a) diode and (b) FET

Compared to the FET, the active layer in the diode is grown on highly doped layer which is different from FET as shown in Fig. 2.9.

Multiplier is basically employed for signal generation at higher frequencies due to superior noise performance and stability. Planar topologies can be realized using diode or transistor-based topologies. Diode-based multiplier are employed for higher order multiplication as transistor-based topologies needs higher power drive and are less efficient. Also added advantage associated with the diode multipliers are its robustness in environmental conditions and insensitive to electro static device-related failures. Most prominent diode multipliers are p-n diodes, varactors, step-recovery diodes and SRD is employed for higher order multiplications due to its inherent strong non-linearity.

Ideally, frequency multiplier should perform the following functions for the input signal as per below table

Input Signal	Output function
Frequency f_0	$N.f_0$
Fractional Frequency $\Delta f/f$	$\Delta f/f$
Phase noise L(f)	L(f) + 20 log N

References

1. D.M Pozar, *Microwave Engineering*, John Wiley & Sons.
2. Steven Maas, *The RF and Microwave Circuit Design Cookbook,* Artech House, 1998.
3. G.J. Wheeler, *Introduction to Microwaves,* Prentice Hall, 1978.
4. Mike Golio, RF and Microwave Semiconductor Device Handbook, CRC Press, 2003.
5. Inder Bahl, Prakash Bhartia, *Microwave Solid State circuit Design*, Second Edition, Wiley Interscience.
6. Alexander Chenakin, *Frequency Synthesizers: Concept to Product*, Artech House 2011.
7. David Q. Xu & C.R. Bramner, "An efficient Technique for Varactor Diode Characterization," *IEEE-MTT Symp. Digest*, pp. 591-94, 1997.
8. K.C. Gupta, Ramesh Garg, Rakesh Chaddha, *Computer-aided design of microwave circuits*, Arthech House 1981.

9. Application Notes, *Harmonic Generation using Step Recovery Diode*, Hewlett Packard.
10. Stephen Hamilton and R. Hall, "Shunt mode Harmonic generator using SRD", *Microwave Journal*, pp. 69-78, April 1967.
11. W.S. Cheung, F.H. Levien, *Microwaves made simple: principles and Applications*, Artech House, 1985.
12. E. Camargo, *Design of FET Frequency Multipliers and Harmonic Oscillators*, Artech House, 1998.

Chapter 3

SRD Multipliers

3.1 Introduction

Schottky barrier, step recovery and varactor diodes are employed for multiplier operations. Schottky barrier is employed for RF to the sub-millimeter ranges, varactor (p-n junction/ Schottky junction) up to several GHz whereas SRD up to 18 GIIz. FET family such as metal epitaxial semi-conductor field effect transistor (MESFET), high electron mobility FET (HEMT), metal oxide FET are employed for mixers apart from amplifiers, oscillators, modulators, switches, etc. at microwave frequencies. Schottky barrier contact is formed due to metal and semi-conductor junction and can be converted to ohmic contact by having metal work function smaller than the semiconductor. So Schottky barrier is employed for rectification whereas simple p-n junction can be employed as varactor (variable capacitance). The I-V relation of the junction can be represented as [1]

$$I(V) = I \exp\left(\frac{qV}{\eta kT}\right)$$

k is boltzmann constant, T is absolute temperature, I is the reverse saturation current

The reverse saturation current effects the forward I/V characteristics as well as *knee voltage*.

The capacitance-voltage (C-V) characteristics can be represented as:

$$C(V) = \frac{C_0}{\sqrt{1 - \frac{V}{\phi}}}$$

C_0 is the junction capacitance at zero voltage, ϕ is the built-in voltage of the junction.

3.2 SRD Multiplier

Signal generation with good stability is one of the most critical problems in communication systems. The most prominent signal generations at microwaves are either DRO or TCXO with multiplier chain. The disadvantage of the DRO is the position accuracy of the external puck along with the repeatability of the results. Also temperature stability of the circuit depends on the DR puck which is poor. In addition to that the puck size is too big for the lower end of the microwave range. In comparison to that the TCXO is most stable but it requires multiplier chain for stable output power. Frequency multipliers provide a convenient means for converting a very low frequency oscillator source to higher frequencies for LO applications. This multiplication can be done using either active device (HEMT, MESFET) multiplier or using passive multipliers (using diode). The disadvantage inherent in active multipliers [especially when the multiplication order is more than 3] is the requirement of high drive power and less efficiency due to square law dependencies. Besides this, they are costlier compared to diode multipliers. The diode multipliers are basically robust in environmental conditions and insensitive to ESD related failures which makes it still attractive to be used in local oscillator chain of receiver/transmitter up to X band [2]. In the family of diodes, SRD is theoretically the most non-linear of all the diodes. This is due to two stable response, namely high impedance and low impedance occurring in a very small time. The ease of implementation and fabrication makes it a viable option. The efficient simulation using CAD tools for SRD multiplier is difficult due to extremely non-linear nature of SRD which fails to converge non-linear harmonic balance analysis. It can be due to improper modeling or onset of parametric instabilities. The present methodology the close proximity achieved in the simulated and measured results with simple modelling approach along with suppression of parametric oscillations [3]. This chapter elaborates the modelling approach, realization of the planar resonant circuit and the complete multiplier design.

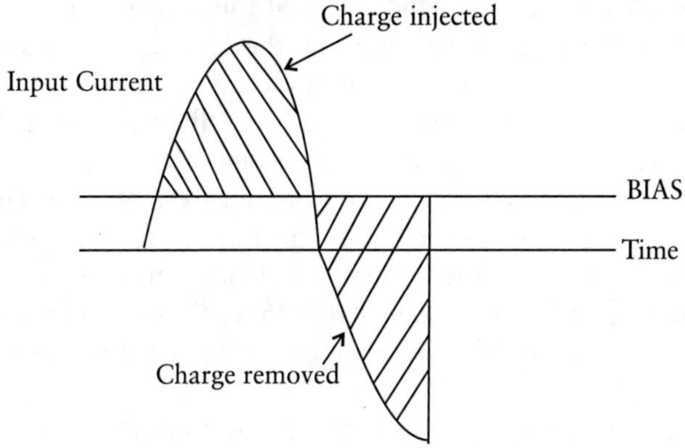

Fig. 3.1: Typical charge behaviour of SRD

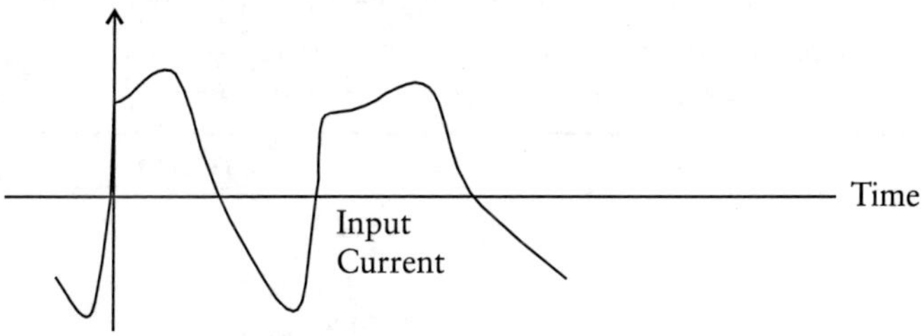

Fig. 3.2: Distorted forward current in SRD

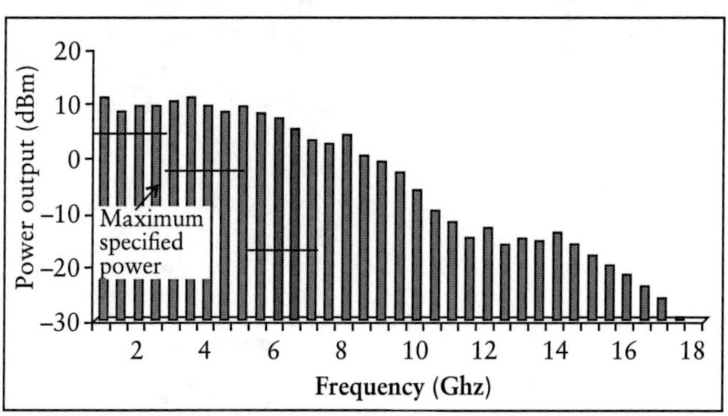

Fig. 3.3: Typical spectrum envelope

3.3 Main Parameters for Diode Selection

Breakdown voltage V_{br}: Determines the maximum amplitude of the impulse and the energy in the impulse.

Reverse bias capacitance C_{vr}: Determine the energy in the impulse as well as impedance level of the ringing line.

Series resistance R_s: Determine the loss that occurs in the ringing line and in the diode input circuit.

Minority carrier lifetime τ: Determines the loss occurs during forward charge storage due to carrier recombination along with bias resistance which develops the diode bias due to rectification current.

Transition time t_t: Determines the diode to form the required pulse width and sets the upper output frequency limit.

Package inductance L_p: Determines the proportion of the energy in the total drive inductance which is not transferred to the ringing circuit.

Thermal resistance θ: Determines the amount of power that can be dissipated in the diode.

Package capacitance C_p: Determines the proportion of overall capacitance and to be lower compared to C_{vr}

Table 3.1: Basic equations for the diode selection

Parameters	Criteria
Transition time	$T \leq \frac{1}{f_0}$
Impedance level	$10\Omega \langle \frac{1}{2\pi f_0 C_{VR}} \langle 20\Omega$
Lifetime	$\tau \langle \langle \frac{1}{2\pi f_{in}}$
Package inductance	$L_p \langle \frac{1}{(2\pi f_0)^2 f_{in}}$
Power handling	$\langle \frac{1}{\frac{1}{0.63\eta} - 1} P_{DISS}$

3.4 Standard Diode Modelling

The diode can be described by a model having fourteen parameters. Out of these, five parameters, namely Is (reverse leakage current), Rs (diode series resistance), N (emission coefficient), Bv (diode breakdown voltage) and Ibv (diode breakdown current) are described by the DC I-V characteristics. The next five parameters, namely Cjo (zero-bias junction capacitance), M (grading coefficient), Vj (junction potential), TT (transit time) and Fc (coefficient for capacitance) describes the non-linear capacitance characteristics of the diode [4]. The rest describes the temperature dependence and noise characteristics of the diode. All parameters are not required to be modeled and depending on the application, a subset of these can be generated. It has been verified that N (ideality factor) has the greater role in mixer operation while M (grading coefficient) in multiplier operation. The non-linear diode model considered for the extraction of the parameters as shown in Fig. 3.4.

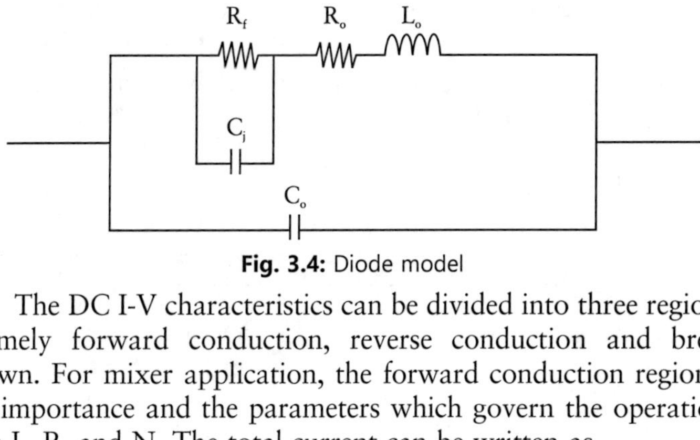

Fig. 3.4: Diode model

The DC I-V characteristics can be divided into three regions, namely forward conduction, reverse conduction and break down. For mixer application, the forward conduction region is of importance and the parameters which govern the operations are I_s, R_s and N. The total current can be written as

$$I_d = I_s\left[\exp\left(\frac{V_d}{N.V_t}-1\right)I_{bv}\exp\left(-\frac{V_d+B_v}{V_t}\right)\right]$$

where Is is the reverse saturation current, N is the ideality factor, V_d is the voltage across the junction, B_v is the breakdown voltage and V_t is the thermal voltage.

$$I_s = qA\left(\frac{D_p P_{no}}{L_p}+\frac{D_n N_{po}}{L_n}\right)$$

where D is the diffusion coefficient and p_n is the concentration of electrons at the n-side.

The model parameters can be derived from the V-ln (I_d) curve and it is represented in Fig. 3.5.

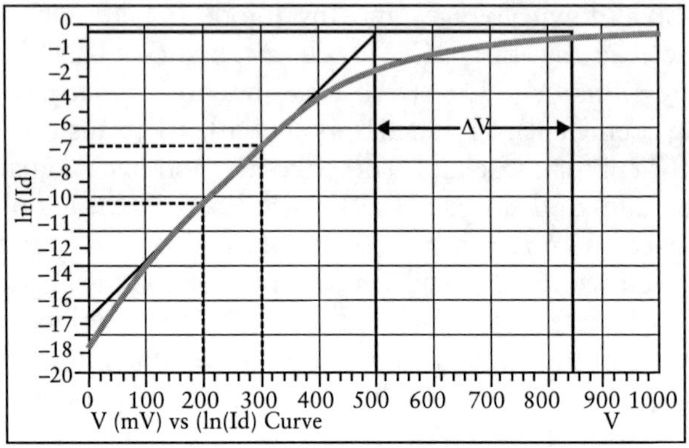

Fig. 3.5: Diode characteristics

The y-axis intercept gives the Is value, the slope N value and R_s is derived indirectly. The values thus obtained were fitted into the model and fine tuned to match the measured I-V characteristics in the forward conduction region. The measured values of I_s, N and R_s are 4.14E-08 A, 1.265 and 0.52 Ω respectively. Both drift and diffusion component has been taken and plotted with the measured value as shown in Fig. 3.6.

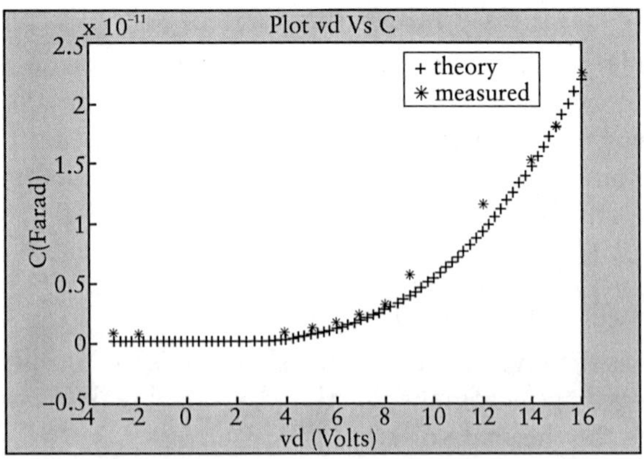

Fig. 3.6: Voltage-current curve

Non-linear Q-V characteristics

$$Q_d = Q_{d,\,dep} + Q_{d,\,diff} \qquad \text{... (3.1)}$$

$$Q_{d,diff} = \tau.\frac{\partial I_d}{\partial V_d}$$

$$Q_{d,dep} = \left[-C_j V_j \frac{C_j V_j}{(1-M)\left(1-\dfrac{V_d}{V_j}\right)^{(1-M)}} \right] \text{for Vd < Fc.Vj}$$

$$= \frac{C_j(1-F_c)^{-M}\left(1+M\left(\dfrac{V_d}{2}-F_c.V_j\right)\right)}{(V_j(1-F_c))V_d + Q_0} \qquad for\,Vd < Fc.Vj$$

$$Q_0 = -C_j V_j (1-F_c)^{-M}\left(\frac{1-F_c}{1-M} + F_c\left(1-\frac{M}{2}\frac{F_c}{1-F_c}\right)\right)$$

where C_j is junction capacitance, F_c is coefficient for forward bias depletion capacitance, M is grading coefficient, TT is transit time.

The non-linear junction capacitance C_j is contributed by depletion and diffusion capacitances.

$$C_{dep} = \frac{C_{j_0}}{\left[1-\left(\dfrac{V_d}{V_j}\right)\right]^M}; \text{Vd < Fc.Vj}$$

where M varies from 1/2 to 1/3.

$$C_{dep} = \frac{C_{j_0}}{[1-F_c]^{1-M}}\left[1-F_c\left(1+M)+M.\frac{V_d}{V_j}\right)\right]; \text{ Vd} \geq \text{ Fc.Vj}$$

$$C_{diff} = \frac{q}{kT}\left[\frac{qL_p P_{no}}{L_p} + \frac{qL_n P_{no}}{2}\right]$$

$$C = C_{dep} + C_{diff} + C_m + C_p$$

$$C_{diff} = \frac{q}{kT}\left[\frac{q.L_p.P_{no}}{L_p} + \frac{q.L_n.P_{no}}{2}\right]e^{\frac{qV}{kT}}$$

So $C = C_{dep}+C_{diff}+C_m+C_p$

C_m is metal capacitance, C_p is poly capacitance.

At higher frequency Cdiff ~$(\omega)^{-1/2}$ so effect is more pronounced at lower frequency.

$$C = \frac{\partial Q_d}{\partial V} \quad \text{... (3.2)}$$

The extracted parameters are fitted back into the above diode equation (3.2) and the DC characteristics are plotted [Fig. 3.7]. It resembles the closed fitting of measured and theoretical modelling. The conversion matrix for the non-linear capacitor can be written as

$$I(t) = \frac{\partial Q}{\partial t} = C(t)\frac{\partial V(t)}{\partial t} + V(t)\frac{\partial C(t)}{\partial t} \quad \text{... (3.3)}$$

which can be expressed as

$$I = j\omega.C.V$$

where ω is the diagonal matrix whose elements are mixing frequencies. As the non-linear device is the series combination of the depeletion layer capacitance and bulk series resistance so the displacement current flowing through the device must be equal to the conduction current flowing through the undepeleted resistive region.

The current in the undepeleted region is

$$I = qN_d v(E)$$

I, represents the drift current where $v(E)=\mu E$

So $I = I_{drift} + I_{diffusion}$

$$I = qN_d v(E) + qD(-\partial n/\partial x) \quad \text{... (3.4)}$$

For a proper diode operation eq. (3.3) should be equal to eq. (3.4).

The non-linear modelling equations describing the conversion loss closely resembles the values obtained by modelling [5].

The non-linear model parameters of the diode are derived based on the measured values. The extracted model parameters are further optimized by comparing the theoretical and measured

I-V and C-V data. Extracted parameters are fitted back into theoretical equation using MATLAB. Modelling approach is validated with successful realization of high frequency circuit.

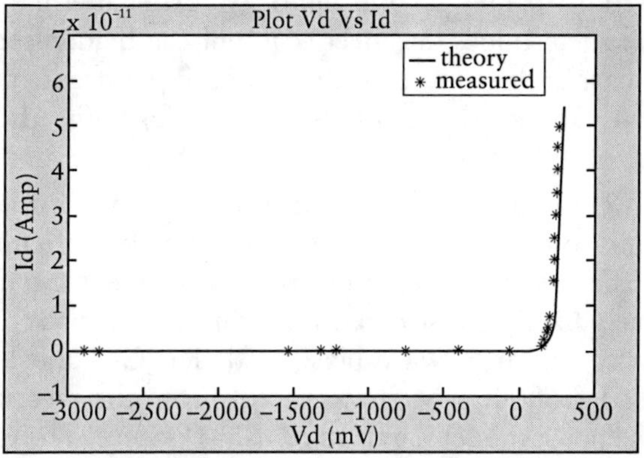

Fig. 3.7: Voltage-capacitance curve

3.4.1 SRD Model

Diode can be described by a model having fourteen parameters. Somc diode parameters are useful to describe the DC I-V characteristics while other set describes the C-V characteristics. The rest describes the temperature dependence and noise characteristics of the diode. All parameters are not required to be modeled and depending on the application, a subset of these can be generated. It has been verified that N (ideality factor) has the greater role in mixer operation while M (grading coefficient) in multiplier operation. Step recovery diodes (SRD) considered as charge controlled switch, are most preferable in multiplier operation [3].

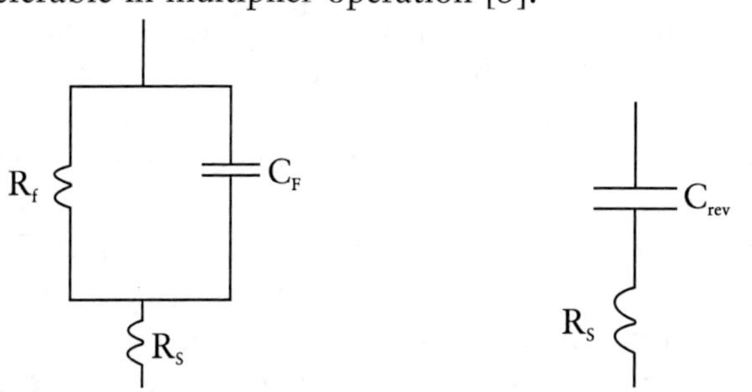

Fig. 3.8: SRD equivalent model

These diodes having the C-V curve that is theoretically the most non-linear of all varactor diodes can be approximated as step junction. They are different from ordinary P-N diode in terms of transit time (T_t) and minority carrier lifetime (τ). The transit time determines the effective impulse at higher frequencies whereas carrier lifetime sets the lower input frequency limit ($\tau \geq 10/f_{in}$). The diode cut-off frequency (f_T) limits the output frequency of operation.

The SRD is assumed to be driven hard into forward and then hard into reverse direction so the equivalent circuit of step recovery diode can be represented by two linear circuits [Fig. 3.8], namely fixed reverse capacitances in reverse bias conditions and short circuit at forward bias [2, 3]. The C-V and I-V curve of the SRD diode has been measured. From the measured DC values, the Spice model is extracted using P-Spice. The extracted parameters are fed back to the simulation. Major SRD parameters are: C_{REV} (reverse bias capacitance), τ(minority carrier lifetime), transition time(Tt), series resistor (R_s), package inductance (L_p) and diode breakdown voltage(V_{br}). Theoretical equations are used to fine tune the model parameters. Also some redundant parameters are optimized using simulated tool, to have close match with the measured value along with easy convergence of solution.

$$\tau = R_F C_F \qquad \text{... (3.5)}$$

3.5 Design Sequence

The multiplier has to be designed for an output frequency of 6.25 GHz (nf_i) with an input of 1.25 GHz (f_i) with minimum input drive for C-band transponder application. The multiplier design consists of a high pass filter followed by a impulse generator circuit [Fig. 3.12, A]. High pass filter avoids unnecessary interaction with the source circuitry [3]. The efficiency of the multiplier depends on the formation of the impulse at the reverse cycle which depends on the stored charge to return to zero at the time when current has the maximum negative value. This can be adjusted by the values of the impulse generator. The impulse generator circuit consists of the inductor in series with the shunt capacitor which is realized in microstrip to convert the input sine wave into periodic pulse [Fig. 3.12, B]. The impulse generator can be followed by a resonator to resonate at the output frequency. The impulse generator terminated in a resistor load gives rise to impulse waveform and associated comb spectrum. Terminations with the resonant network give rise to peaking at the output frequency by concentrating the energy at the output frequency. The resonant structure is viewed as the first section of the output filter of the multiplier. Basically, a quarter wavelength line with a low value tunable capacitor is chosen as a resonant network which results in high quality factor ($Q \approx \Pi n/2$). But this approach gives rise to repeatability problem along with the unrealistic range of values. The present methodology reports the design based on two quarter wavelengths transforming low to high and again high to low impedances which is followed by the band pass filter. The critical portion is the electric phase between the resonator and the filter. The band pass filter topology is selected to be hairpin to have compactness of the circuit. The number of resonators chosen is on the basis of the bandwidth and cut off attenuation rate which can be further optimized to have better suppression of harmonics of the multiplier. The self bias can be realized using high impedance line terminated with sensistor (10 K ohm) and optimized to remove parametric instabilities by incorporating capacitors at appropriate locations. Self-bias advantage is the low cost, temperature compensation and self-adjustment for different power levels.

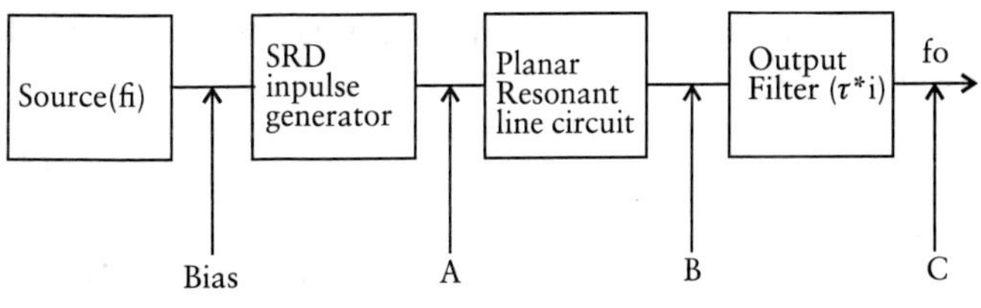

Fig. 3.9: The block schematic of an SRD multiplier

Major criteria to be ensured for proper operation of SRD multiplier are:

1. Breakdown voltage (V_{br}) must be greater than minimum reverse voltage across the diode
2. Carrier lifetime of the diode (τ) should meet the criteria of $\tau \geq 10/f_{in,}$ where f_{in} is the input frequency.

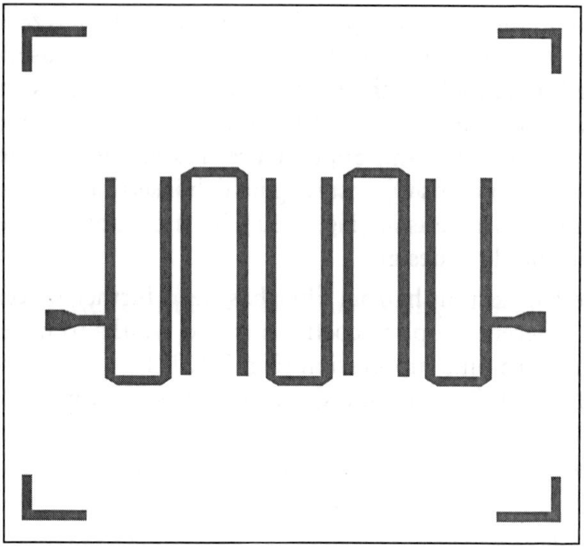

Fig. 3.10: Hairpin standard topology

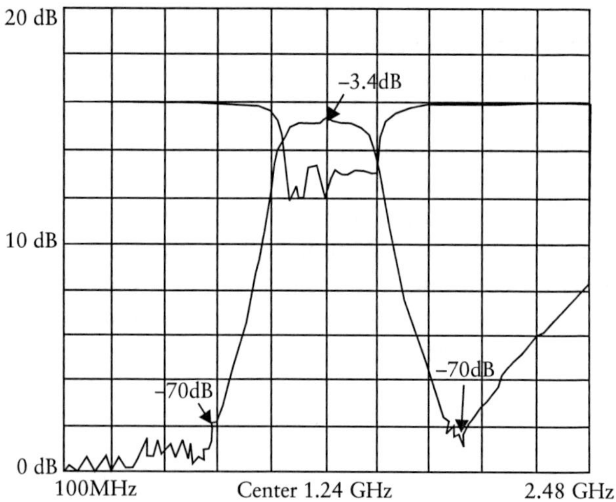

Fig. 3.11: Measured results

The other way round in this approach is to make strip width as wide as possible to maximize the Q_u. But as the strip width becomes wider, the first section gap becomes smaller and put hindrance on the practical limitation of the realizability [9-10].

$$\text{Insertion Loss (dB)} = \frac{4.343 f_0}{B.W \times Q_U} \sum g_i \quad \text{... (3.13)}$$

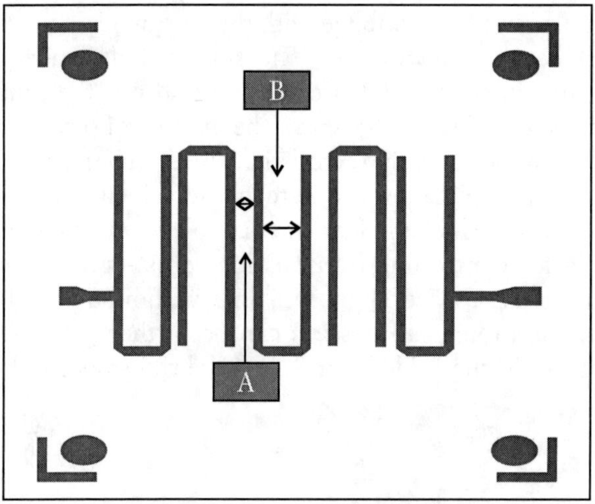

Fig. 3.12: Hairpin modified topology

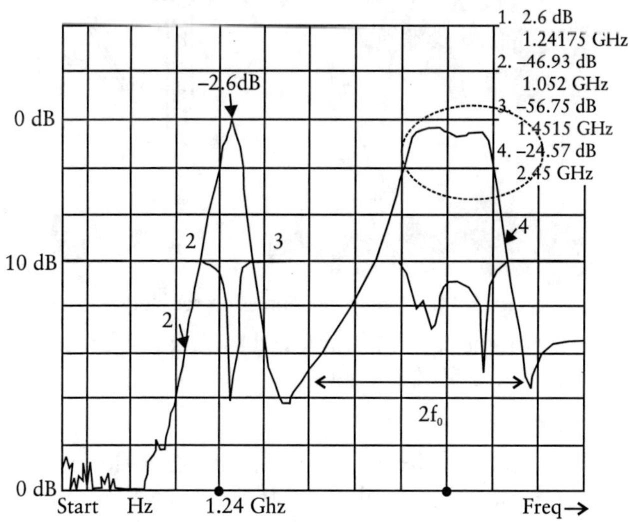

Fig. 3.13: Measured results

The measured results shows improved characteristics along with better insertion loss, which validates the theoretical approach.

3.5.2 Harmonic Suppression Topology

The proposed structure giving harmonic suppression along with steep attenuation rate is shown in Fig. 3.14. The concept of edge coupled has been clubbed with the hairpin giving a compact size and better performance. The standard design equations as described in section 3.5 are used to find out the number of sections and coupling coefficients. The number of sections chosen is six for symmetrical characteristic. The elimination of second harmonics can be achieved due to phase velocity equalization. Also the coupling coefficient and coupled lengths have been optimized for controlling transmission zeros. Further the input and output planes are aligned in line without much bending [11]. The harmonic suppression can be further improved with via insertion in the folded portion of the transmission line.

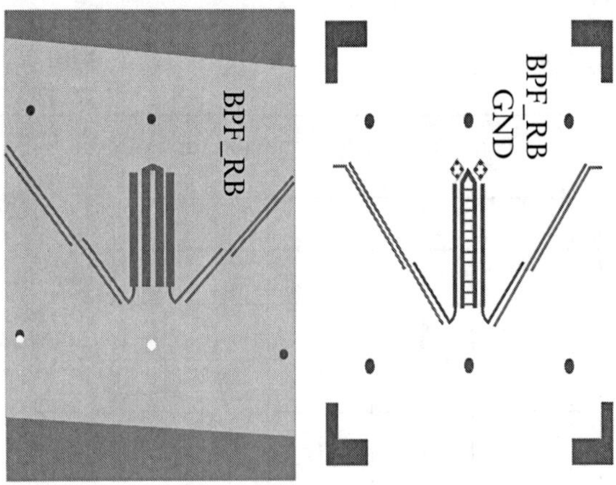

Fig. 3.14: Harmonic suppression topology

These topologies can be employed to suppress the harmonics but at the expense of the bigger size.

3.5.3 Measurement Results

All the filter structures are simulated using electromagnetic tool for optimizing the coupling coefficient as well as minimizing cross coupling. All the filters are implemented on RT Duroid dielectric substrate having permittivity of 10.5 and thickness of 50 mils. The modified filters are designed at 1.24 GHz with bandwidth of 10%. The measured results of modified hairpin topologies are shown in Fig. 3.15. The insertion loss is better than 2.2 dB. Tapering is done for better alignment with subsequent circuit giving .5 dB extra loss. The return loss is better than 18 dB in both the cases. The size of the hairpin cards shown in Fig. 3.12 is 50 mm × 32 mm. The harmonics are suppressed better than 20 dB in the proposed filters. The extra insertion loss compared to simulate one is due to conductor and radiation loss. The dimensions are 93 mm × 60 mm which can be reduced further by 20 mm.

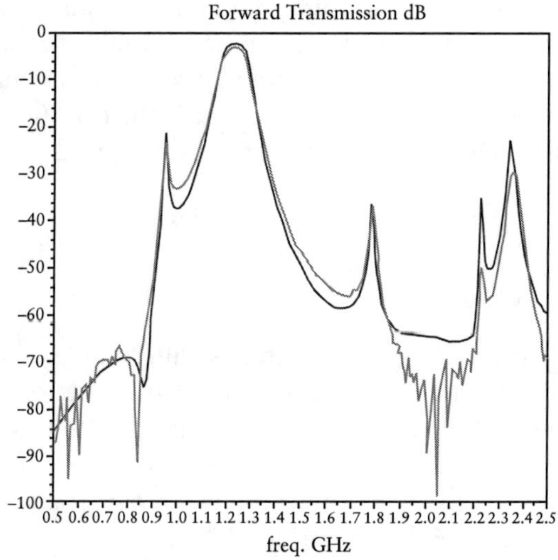

Fig. 3.15: Simulated and measured results

Presented hairpin filters having improved characteristics over conventional structures using standard design equations. The line widths and spacing can be easily etched using standard fabrication techniques. Narrower bandwidths have been achieved without additional components. The structures are validated having close match between the simulated and practical results. This approach can be extended to much higher frequency range without compromising the filter performance.

3.6 Simulation Set-Up

In order to perform non-linear harmonic balance analysis, the knowledge of diode model parameters are essential and the accuracy of determining these will translate into a close match between the simulated and measured results. The full circuit is integrated and simulated using harmonic balance tool (Fig. 3.16). The model parameters of SRD are incorporated and circuit is optimized for efficient output power at n × f_0 and for suppression of undesired harmonics. The full circuit is realized on the soft substrate having dielectric constant of ε_r=2.55, with the thickness of 50 mils. The input power of minimum 17.5 dBm is found out to be optimum for the efficient multiplier operation. For better temperature stability and proper bias generation a quarter wavelength line terminated with the 10 kΩ sensitor is incorporated. The value is calculated using Eq. (3.14).

$$R_B \approx \frac{2\tau}{\left[\Pi\left(\frac{f_o}{f_{in}}\right)^2 C_R\right]}\Omega \qquad \ldots (3.14)$$

The input consists of the high pass filter implemented using high impedance shorted stub and lumped capacitor, matching circuit is implemented using open circuit stub followed by the impulse generator circuit realized in microstrip configuration at fi. The output consists of the resonant circuit followed by the band pass filter at 6.25 GHz (Fig. 3.16).

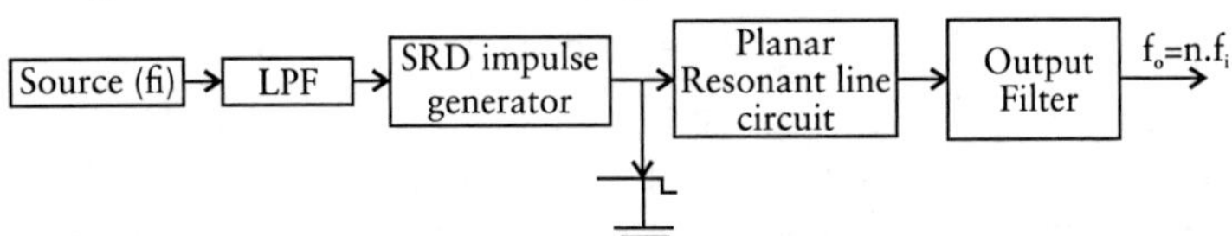

Fig. 3.16: SRD-based multiplier

The resonant circuit is designed using two λ/4 sections with further incorporation of transmission line having required phase shift to achieve maximum power transfer and followed by five sections of hairpin filter.

Simulated results as shown in Figs. 3.17-3.20 show phenomena like impulse generation, damping and overall output spectrum.

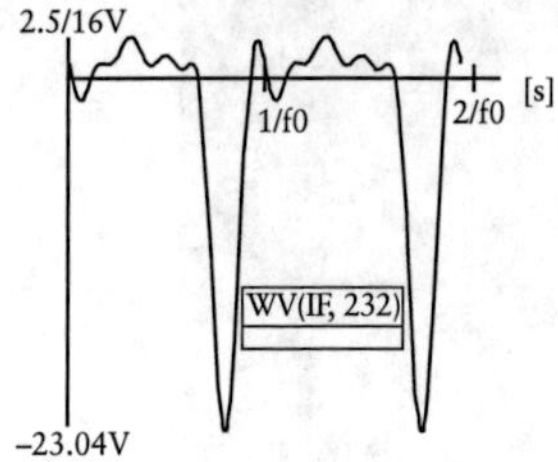

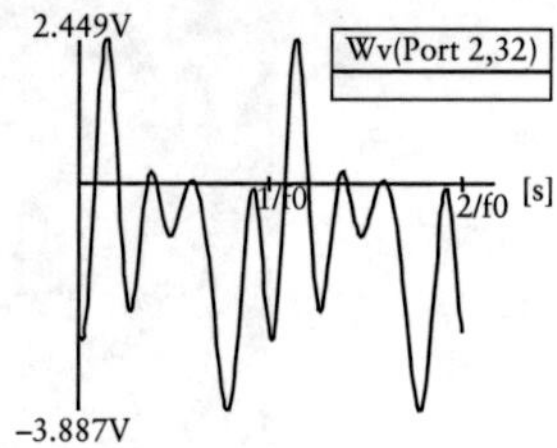

Fig. 3.17: Simulated impulse generation **Fig. 3.18:** Simulated damped waveform

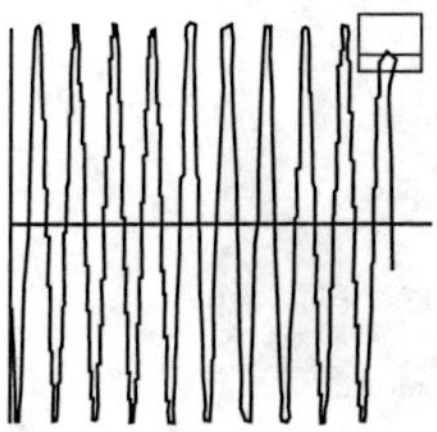

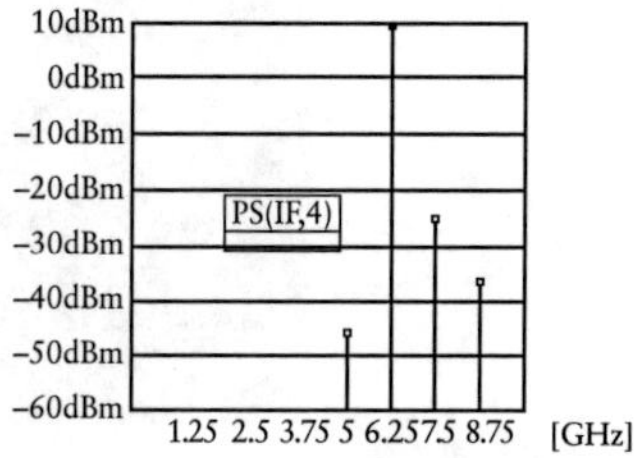

Fig. 3.19: Simulated output waveform **Fig. 3.20:** Simulated output spectrum

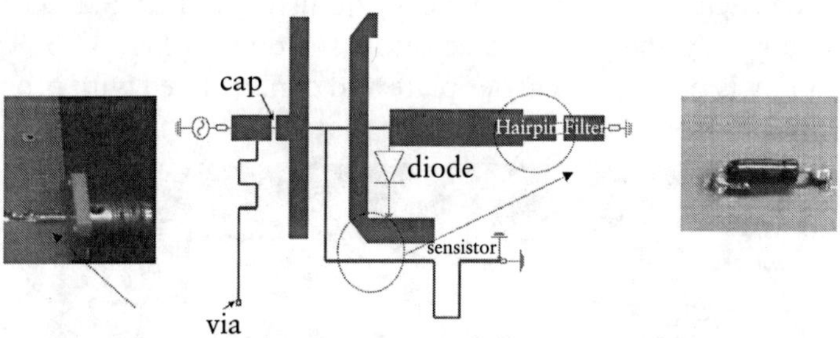

Fig. 3.21: Layout of the designed SRD multiplier

Via is used in the input high impedance section acting as a high pass filte as shown in Fig. 3.21r. Via is modelled using the below formulae

$$X_L = jwL,\ L = 7.9577\ X_L/f(\text{GHz})\ \text{nH} \qquad \ldots (3.15)$$

3.7 Assembly and Testing

SRD assembly to have proper connection is important so as to get optimum performance from the circuit. Ring having silver-plated copper wire is employed for properly arresting the diode.

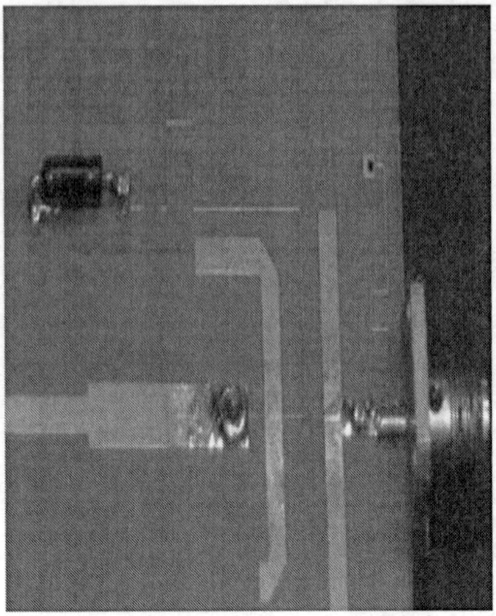

Fig. 3.22: Assembling of the SRD

Realized diode-based frequency quintripler (1.25/6.25 GHz) incorporating the SRD and sensitor is shown in Fig. 3.23. The assembly is put on the carrier plate and conectorized with proper mounting provisions in the carrier plate.

Fig. 3.23: Realized SRD multiplier

3.7.1 Testing Methodology

Input signal frequency	:	1250 MHz
Input signal level	:	+18 dBm
Output signal frequency	:	6250 MHz
Output signal level	:	+6.5 dBm
Input and output impedance	:	50 Ω
Return loss	:	>12 dB

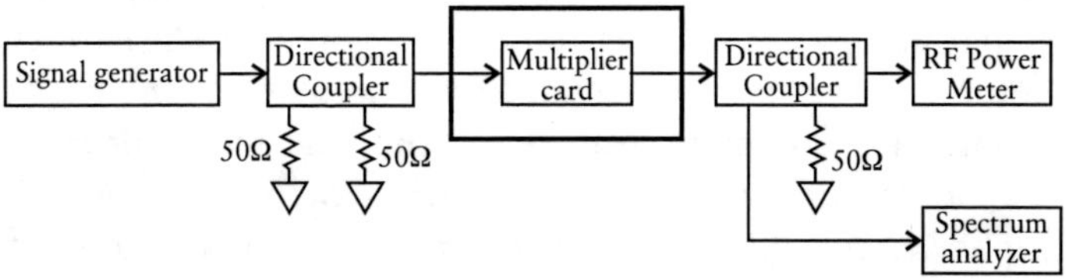

Fig. 3.24: Test set-up for multiplier card

The output power of 7.0 dBm at 6.25 GHz comes with the input drive of 18 dBm at 1.25 GHz having 1-dB bandwidth of 150 MHz. The hairpin topology gives the insertion loss of 2.5 dB with the return loss of better than 13 dB and having the 3-dB bandwidth of 250 MHz [Fig. 3.25].

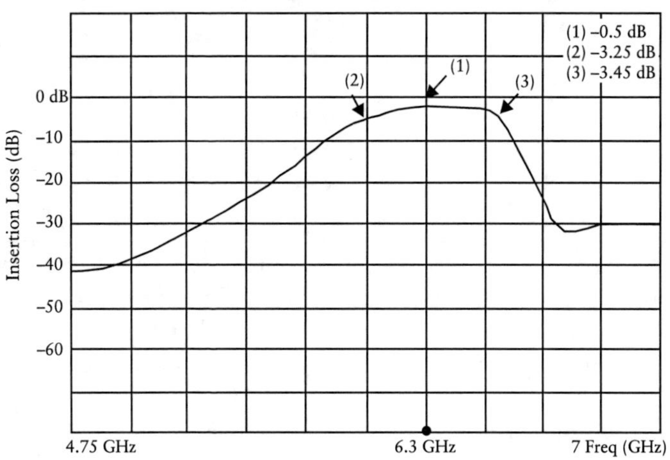

Fig. 3.25: Realized structure

The resonators consist of two quarter wave transformers at the output connected with filter with the proper phase matching. The harmonic suppression of better than 30 dB is achieved. The multiplier is off tuned by 50 MHz which relates with fabrication inaccuracies changes in permittivity of the substrate and can be taken care intolerance analysis. The main advantage of the present topology is the elimination of the lumped element at the output. Slight deviation from simulated output power is attributed to the leakage of the signal at the bias input. Also hairpin topology gives slightly more insertion loss. The extracted model can be used in commercial circuit simulation for designing SRD circuits The main feature of the selected approach is the demonstration of circuit simulation of highly non-linear multiplier circuit. SRD multiplier is reported to be suitable for output between 5 GHz-20 GHz and up to 10th harmonic only. The input power is varied from 18-23 dBm and corresponding voltage and output power are tabulated as shown in Table 3.2.

Table 3.2: Variation of bias voltage and output power

Pin (dBm)	Bias Voltage (mV)	Pout (dBm)
18	–50.43	3.75
19	–58.8	5.92
20	–188	9.97
21	–283	11.47
22	–378	12.71
23	–494	13.82
24	–622	14.87

At particular power level it is noted that linear increasing the power yields higher output power. The correlation of the corresponding power and voltages show that bias also shoots up in that range. SRD is prone to environmental conditions so thermal variations are performed to find out overall impact on the output power level. The multiplier is also sensitive to temperature variation as the operating frequency is high and multiplication factor is also high.

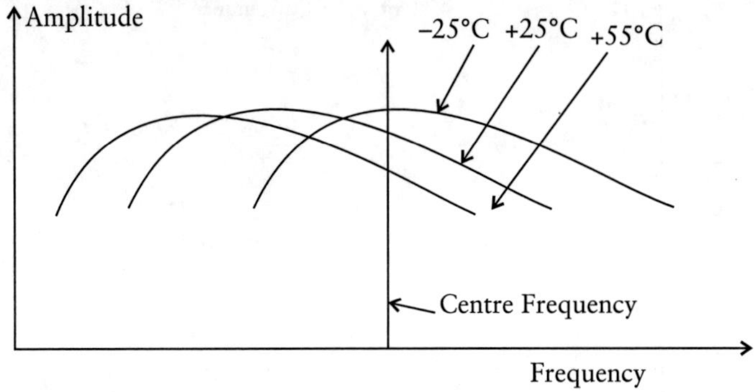

Fig. 3.26: Standard SRD multiplier performance with temperature

Output power is function of input power, centre frequency and operating temperature, so there is a need to optimise the variation within certain limit by varying the tunable elements.

Further the SRD is operated beyond the optimum power so as to achieve oscillations free operations with minimum losses.

Summarization of the overall design of SRD multiplier from L to C-Band.(five times)

- The parameters of SRD are extracted by using measured I-V and C-V data and fitted into P-Spice as shown in Fig. 3.26. Empirical-based approach for the development of the compact model with the fitting of the I-V, C-V characteristics defined by the closed set of equations resulting in the close match between the predicted and measured behaviour of the device. This approach results in the scalability and flexibility.
- Designed multiplier includes no external bias and no tuning elements.
- The measured results of the multiplier is the output power of 6.5 dBm for 18 dBm input at 6.25 GHz with harmonic suppression of better than 30 dB.

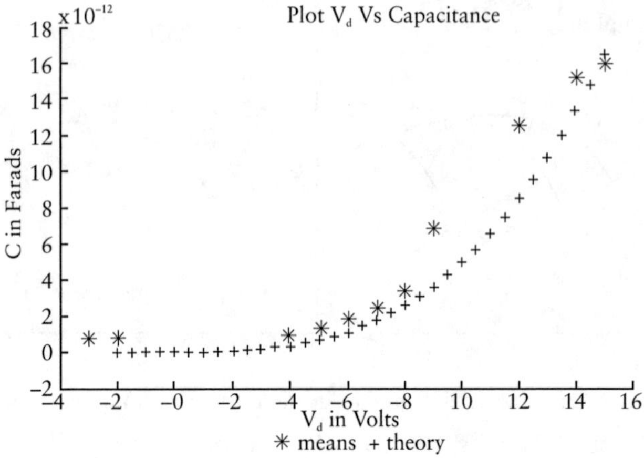

Fig. 3.27: Fitted parameters with the measured value

SRD-based multiplier realization can be segregated into various parts:

1. Diode biasing circuits and matched filters
2. Self-bias approach employs sensitor
3. Impedance matching circuitry

SRD is reactive device and circuit avoid incorporation of idlers due to inherent strong non-linearity. Compared to varactors non-linearity which exist due to associated depletion capacitance, SRD uses diode diffusion capacitance for storing charges. This results in closely spaced harmonics but bottleneck of high sensitivity exists in the circuit [13]. So the implementation methodology for design changes with the frequency band. The non-linear model parameters of the diode are derived based on the measured values. The extracted model parameters are further optimized by comparing the theoretical and measured I-V and C-V data.

The main limitation of SRD-based multiplier is the inherent non-linearties resulting in batch to batch variation in the device. The signal source is followed by low pass filter in the design to avoid unnecessary interactions and to improve stability. It is realized at low end of frequency employing lumped inductor and capacitor. The impulse generator circuit consists of the inductor in series with the shunt capacitor followed by resonator circuit to resonate at the output frequency. The realization of these circuits is carried out employing tunable capacitors which caters for the proper matching and operation of the multiplier chain. Sensistor for self-bias is also incorporated with capacitor to remove parametric instabilities associated with this circuit and slight change in power results in degraded performance. Keeping into consideration of the above tunablity option, the circuit needs extensive optimization at card and package level to achieve the desired results.

References

1. Mike Golio, *RF and Microwave Semiconductor Device Handbook*, CRC Press, 2003.
2. Inder Bahl, Prakash Bhartia, *Microwave Solid State circuit Design*, Second Edition, Wiley Interscience.
3. Application Notes, *Harmonic Generation using Step Recovery Diode*, Hewlett Packard.
4. Jian Zhang and Antti Raisanen, "A new model of step recovery diode for CAD", *IEEE MTT-Symp. Digest*, pp. 1459-62, 1995.
5. Kamaljeet Singh and R. Ramasubramanian, "A simple diode modelling approach for mixer applications", *IEEE-APMC-Proc.*, China, 2005.
6. Kamaljeet Singh, Novel Design in Planar Active and Passive Circuits at Microwave Frequencies, PhD Thesis 2010.
7. G.L. Matthaei, L. Young and E.M.T Jones, *Microwave Filters, Impedance-Matching Networks & Coupling Structures*, Artech House, Dedham, MA, 1980.
8. C.M. Tsai and H.M. Lee, "The Effects of component Q distribution on Microwave filters," *IEEE MTT-S*, Vol. 54, No. 4, April 2006, pp. 1545-53.
9. Joseph S. Wong, "Microstrip Tapped-Line Filter Design", *IEEE Trans. MTT*-27, 1, 1979, 44-50.
10. Jia-Shen, G. Hong, M.J. Lancaster, *Micro Strip Filters for RF/ Microwave Applications* IEE Press.
11. Wang, Chang et al., "Miniaturized Spurious Pass band Suppression micro strip filter using Meandered Parallel Coupled lines", *IEE Trans. MTT*, Feb. 2005, pp 747-53.
12. Daniel G. Swansen and W.J.R. Hoefer, *Microwave Circuit Modeling Using Electro-magnetic Field Simulation*, Artech House, 2003.
13. Application Notes (920, 989), *Harmonic generation using step recovery diode*, Hewlett Packard.

Chapter 4

FET Multipliers

Step recover diodes can supply the needed power directly but generally require RF drive power levels that are incompatible for the application requiring low currents and voltages. Development of the compact model with the fitting of the I-V, C-V characteristics defined by the closed set of equations resulting in the close match between the predicted and measured behaviour of the device. This approach results in the scalability and flexibility.

4.1 Introduction

Multipliers based on transistors employ both BJT and FET. They are based on the general principle that conduction takes place during portion of the input sinusoid signal. The DC bias in the circuit is chosen so as to select appropriately highlight the desired harmonics[1]. Compared to diode based multipliers, the basic advantage associated with the transistor based multiplier is its high efficiency due to amplification as well as non-linear behaviour.

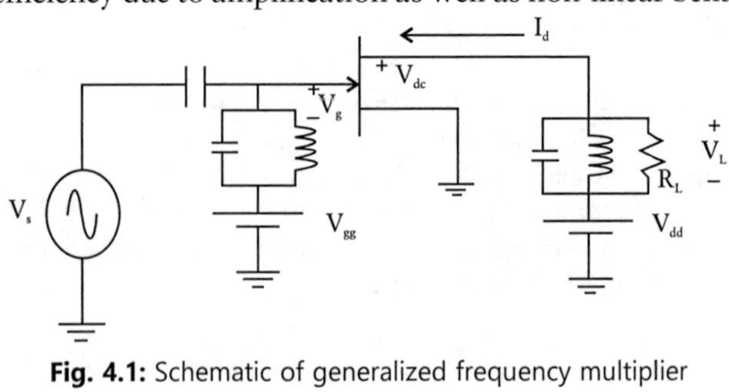

Fig. 4.1: Schematic of generalized frequency multiplier

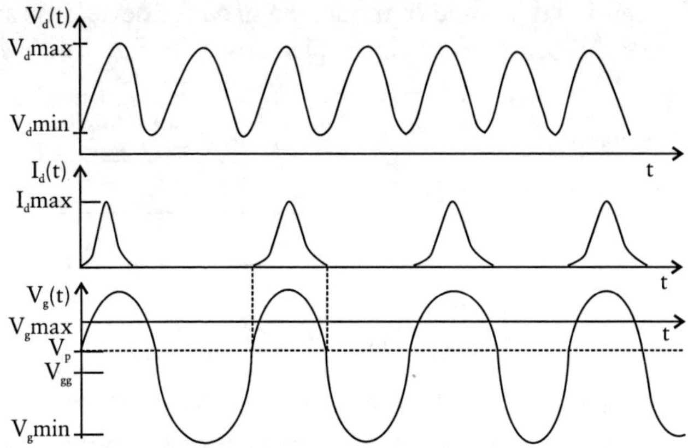

Fig. 4.2: Voltage and current waveform for FET multiplier

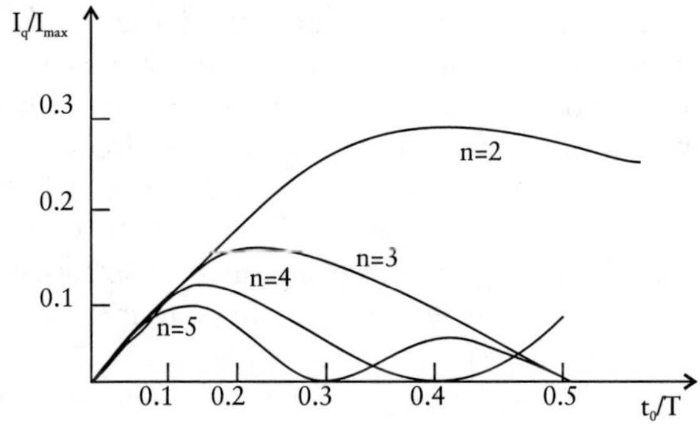

Fig. 4.3: Harmonic current components

The major criteria for the efficient tripler operation are the choice of bias along with the harmonic current feedback to the device. To get asymmetrical output waveform rich in third harmonic the approach adopted in this chapter is by biasing the device in class A and presenting short circuits at the even harmonics and open circuits at the odd harmonics, except for the third harmonic which is matched for maximum output power [2]. This voltage distortion approach provides higher odd harmonic current, compared to the other approaches as described by Thomas and Branner [3]. The major constraint in this approach is placing quarter wavelength transformers for short circuiting the even harmonics. This increases the circuit complexity as impedance at other harmonics also has to be taken into taken into account [4] and should be tuned out.

Conventional method of modelling of active devices is carried out as per the below flow chart, [1] shown in Fig. 4.4.

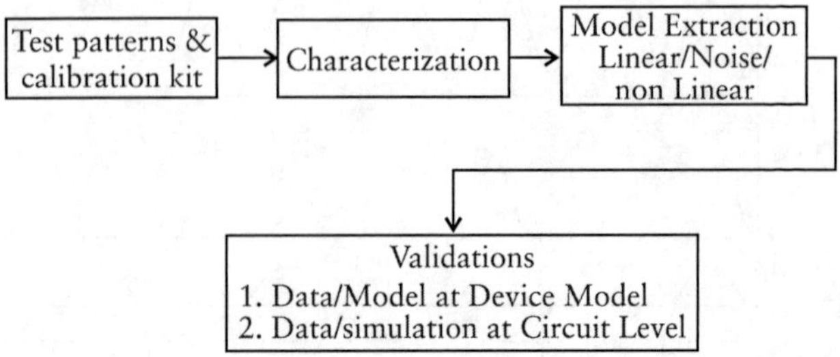

Fig. 4.4: Generic steps for modelling of active devices

In the present methodology approach of harmonic loading the output filter to have a simple topology along with an efficient conversion loss at Ku Band [5] provide desired performance with simple implementation. The major drawback of FET multiplier is the requirement of high power drive to achieve low sensitivity to device parameter variations. The circuits presented are having optimum performance with a lower drive due to harmonic loaded filter approach. There is slight difference in the way a model is implemented in commercial softwares. These differences can be reconciled by mapping parameters from one simulator to another, reorganizing default values of parameters and analyzing their effects as well as modifying the simulator schematic. This approach makes the simulation more near to practical data in the different circuit simulators and has been validated by the development of active multipliers giving close match between simulated and practical results. An active tripler has been designed and developed at Ku-Band based on the non-linear Materka model of HEMT device. The design approach is to generate non-linear model of the active device based on the linear model and various parameter measurements along with validating the same by designing sampler circuits. The accuracy of the model is further enhanced by fine tuning selective parameters (those specifically playing a role in the multiplier design) and finalizing the non-linear model for tripler design. The approach has been validated by the development of active tripler and its characteristics.

4.2 Harmonic Generation in FET Tripler

A multiplier operation depends on frequency of operation, level of input drive, bias conditions and terminations at the fundamental frequency [6-7]. The transconductance g_m directly contributes to device power performance and multiplication gain, whereas f_T (transition frequency) and f_{max} (maximum frequency of operation) gives insight into the impedance of the parasitics elements associated with a given device and how well it will fit to the chosen model. The f_T and f_{max} are given as:

$$f_T = gm/(2\pi Cgs) \text{ and } f_{max} = \frac{f_T}{2}\sqrt{\frac{R_{ds}}{R_g + R_i}} \quad \text{... (4.1)}$$

The standard procedure to design a frequency multiplier is to determine the appropriate biasing and the best conduction angle for the active device to generate a signal waveform rich with the desired harmonics at the output. Major design equations are

$$Ip = Idss\ (1 - Vgs/Vp)$$

Idn = .54 Ip/n for n>1 where n is the harmonic number

$$\text{Conduction angle } \Phi_0 = \sin^{-1}(2Vg/|Vp|) \quad \text{... (4.2)}$$

where $Vgs(t) = V_{G0} + Vg \cos\omega t$

It is known that symmetrical trapezoid at the output is rich in odd harmonics. For symmetrical trapezoid to be rich in third harmonic the conduction angle should be near to $\pi/4$. So from Eq. (4.2) the voltage value at gate will be $Vg = .707|Vp|$.

A multiplier operation depends on frequency of operation, level of input drive, bias conditions and terminations at the fundamental frequency. Third harmonic can be generated with multiple bias schemes. The current waveform rich in third harmonics is generated by biasing the device in class C with an optimum angle of 80°. But this mode of operation requires high breakdown voltage device. The second approach consists of biasing the device in class A. Then further overdriving it until there is clipping due to pinch off on the negative swing and due to gate conduction on the positive swing. In this case, drain current waveform tends to be square for which only the odd harmonics exist. But overdriven class A operation needs a high gate resistance to limit the gate DC conduction for assuring a reliable operation. It has been reported [8] that class B operation generates third harmonic content but optimum conversion loss comes out at much higher drive. Other option to get asymmetrical voltage waveform rich in third harmonic is by biasing the device in class A and presenting short circuits at the even harmonics and open circuits at the odd harmonics, except for the output third harmonic which is matched for maximum output power. This voltage distortion approach provides higher odd harmonic current, compared to the other approaches as described by Camargo [4].

Parameters such as delay time, gate fringe capacitance, source resistance are also important and to be considered for accurate modelling. The parameter delay causes performance issue and can be expressed as

$$\text{Delay time } \tau_T = \frac{1}{2\pi f_T}$$

where $t_T = t_p + t_{fringe} + t_{channel} + t_{transit} + t_{drain}$

The charging time for the pad is given as

$$t_p = \frac{C_p}{g_m W}$$

C_p - pad capacitance; t_p - charging time; g_m - extrinsic transconductance per unit gate width

W - width of the device

Increasing the value of transconductance as well as gate width will decrease the pad charging time. Gate fringe capacitance

$$t_{fringe} = \frac{C_{fringe}}{g_{m_0}} \quad \text{where} \quad g_m = \frac{g_{m_0}}{1 + g_{m_0}.R_s}$$

R_s-source resistance

$$t_{transit} = L_g / v_{sat}$$

The parameter t_{drain} becomes important in case of high drain bias voltage is applied which creates drain depletion region which is on the order of gate length of the device. The voltage gain of the device can be represented as g_m/g_{ds} and measures the short channel effects in the device. The out-put power can be mathematically presented as

$$P_{out} = \frac{1}{8}(I_{max}).(BV_{gd} - V_k)$$

$$= \left(\frac{1}{8}\right)\left(\frac{V_{ss} - V_k}{R_L}\right)^2$$

$$PAE = \alpha\left(\frac{V_{DD} - V_K}{V_{DD}}\right)\left(\frac{1}{G_a}\right)$$

α = 1/2 (for class A)

4.3 HEMT Modelling

Non-linear circuits simulator must have non-linear model which fits the widest range of biases, frequencies and applications as possible. Although there are number of GaAs MESFET models, none has proven to be fully satisfactorily for accurate simulation of the non-linear GaAs MESFET operations. An HEMT model is primarily a GaAs MESFET model where transit time effects have to be characterized more accurately. This is done with the proper description of the drain-source current and the dispersion current[9]. This approach is implemented for Materka model. There is always a trade off when developing a non-linear model like power and noise figure [10]. An effective and practical large signal microwave GaAs MESFET model should satisfy the criteria like accurate representation of the measured DC I-V characteristics and assuring convergence to steady state conditions, accurate representation of all the operating point dependent non-linearties that occurs in large signal operation and supported by a practical characterization method. In this paper the approach was to fine tune the model parameters effectively for tripler and small signal amplifier operations. The major parameters govern the multiplier and amplifier operation is trans-conductance and output conductance. So accurate representation of these non-linear parameter are essential for achieving close analogy between the simulated and measured results. The Materka model is chosen as it gives more accurate representation of saturation of the output power along with the breakdown effects. The experimental characterization and modelling is validated with a close match between the simulated and measured results for multipliers at two different frequencies.

Various inherent advantages and disadvantages of models are reported in the literature as no model completely applicable for all non-linear circuits[11]. The Curtice Quadratic model [12] was the first one introduced for the simulation of non-linear microwave FET devices. Afterwards curtice cubic, statz-pucel, TOM and Materka were developed. Due to better description of the saturated output power Materka-Kacprzak was chosen in the presented approach [4]. Non-linear elements are C_{GS} (gate-source capacitance), D_{GD} (gate-drain diode), D_{GS} (gate-source diode) and I_{DS} (drain-source current) where I_{DS} is given as:

$$I_{DS} = I_{DSS}(1 - V_{GS}/V_P)^2 \tanh[(\alpha_M Vds/(Vgs–Vp)] \qquad \text{... (4.3)}$$

where α_M is drain saturation voltage, Vp is pinch off voltage

The presented approach assumes non-linear characteristics of HEMT model as similar to MESFET. The equivalent circuit can be separated out as an extrinsic and intrinsic part representing HEMT. The extrinsic part is related to device layout and modelled with linear element. The intrinsic part represented by non-linear currents and charge generated conduction and displacement currents. DC I-V curves for different transistors are where taken AC parameters were found out by overall measured performance of the sampler circuits along with fine tuning of some dummy parameters for convergence improvement [5].

4.3.1 Modelling Approach

The Curtice Quadratic model was the first one introduced for the simulation of non-linear microwave FET devices. But this model lacks to predict inter-modulation distortion, as there is no third order term in the drain current description. Curtice cubic overcomes the above-mentioned hindrance but cubic relation provides parameters resulting in the pinch off either making current zero or transconductance zero but not both simultaneously. Statz-Pucel model predicts I_{ds}-V_{gs} as linear and quadratic for different operating voltages but drawback is the consideration of the drain current as independent of drain voltage. Also transistor output resistance Rds is not correctly represented. TOM model corrected the anomaly in the pinch off voltage description but lacks simplicity. Materka-Kacprzak developed a model which gives accurate description of saturated output power. Non-linear elements are C_{GS}(gate-source capacitance), D_{GD}(gate-drain diode), D_{GS}(gate-source diode) and I_{DS}(drain-source current).

The present chapter assumed non-linear characteristics of HEMT model as similar to MESFET [Fig. 4.4]. The equivalent circuit can be separated out as an extrinsic and intrinsic part representing HEMT. The extrinsic part is related to device layout and modelled with linear element. The intrinsic part represented by non-linear currents and charge generated conduction and displacement currents.

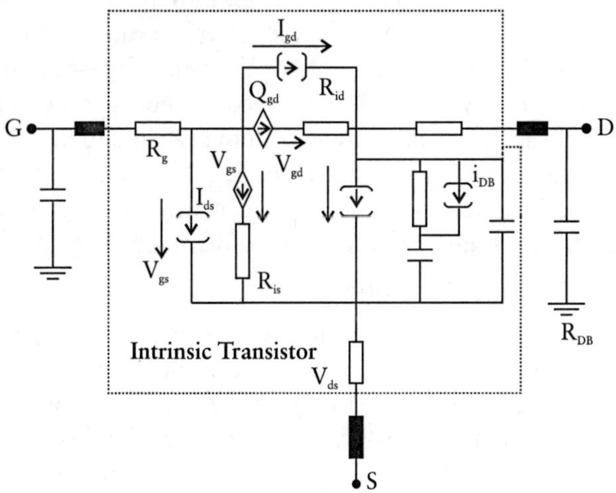

Fig. 4.5: HEMT model

4.3.2 Non-linear Model Extraction

Modelling process includes three steps: Characterization, parameter extraction and modelling [1]. Accuracy of any device model depends upon the accuracy by which the model parameters are determined. Data-based approach (also known as measurement based) has been used having series of DC and S-parameter measurement has been performed over a large combination of gate and drain characteristics. The DC data of five transistors are measured (to take into account for device to device variation) which gives first order estimation of device performance characteristics. Figure 4.4 shows the measures I-V plot of one of the devices. The major DC parameters effecting I-V curve are found out to be V_{TO}, Alpha (α), Beta (β), R_g and R_s. RF characterization and parameter extraction is done with the selection of equivalent circuit topology (in present case 'Materka') which is estimated using combination of DC parameter extraction and RF measurement. Large signal characterization is accomplished using either load pull or by large signal S-parameter extraction. Presented methodology describe latter approach where measurement of gain compression, saturated power, efficiency and harmonic distortion of the sampler circuits like amplifiers and triplers (section 4.3.2, 6.2 are carried out at Ku-band). With the above-measured values the AC parameters of the model, namely Gamma(characterizing Vds dependence on the transconductance), τ(characterizing transit time delay), C_{ds}, C_{gs} have been optimized. The parasitic has been substituted by equivalent-lumped elements to account for accurate characterization. Model is than revisited in line with the measured data and main parameters V_{TOAC} (zero bias threshold), C_{DSO} (output capacitance), $G_{AMMAXAC}$ (peak transconductance), are fine tuned to have steady state convergence.

4.3.3 Experimental Verification

The experimental verification of the presented approach is carried out by designing amplifiers and multipliers at the higher frequencies [13]. The measurements were taken for different drain and gate biases. The device is operated in class A having Vds = 1.5 V and Ids = 12 mA. The CAD tool for designing is employed and the etched cards are attached on the 10 mils alumina kovar plate. Measured results are plotted and shown in Fig. 4.6 and Fig. 4.7.

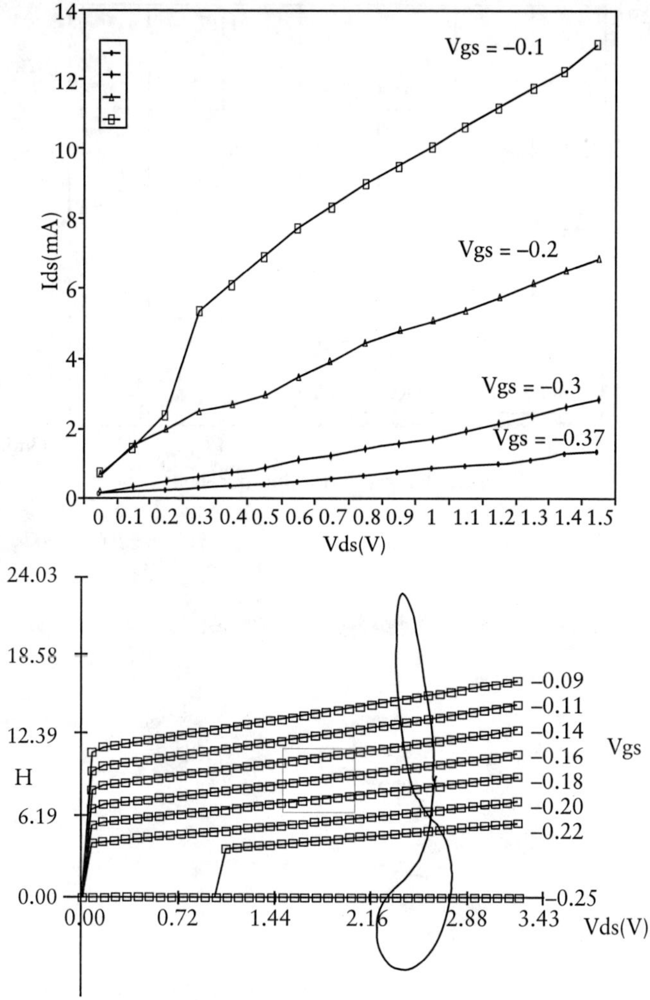

Fig. 4.6: Measured I-V plot of HEMT device

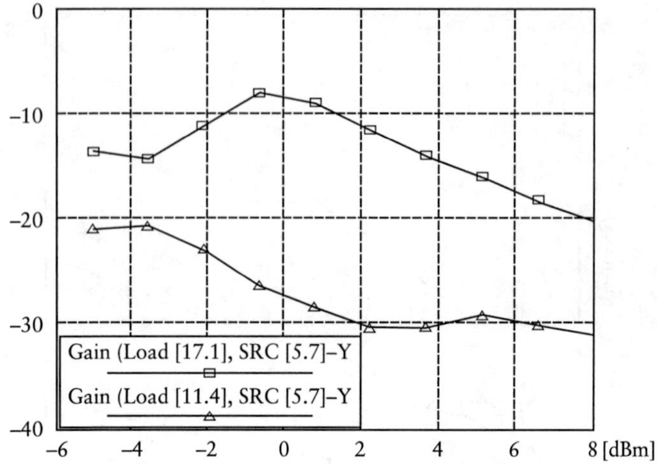

Fig. 4.7: Ellipsoid indicating non-linearity

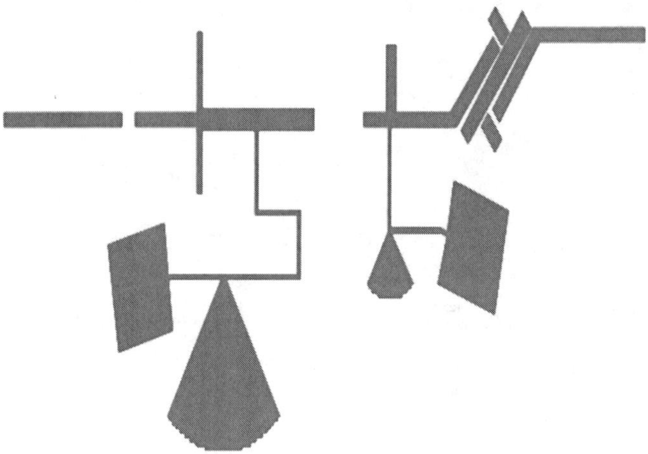

Fig. 4.8: Complete tripler (a) Layout (b) Simulated conversion loss

In summarization realization of 'Materka model' for HEMT device using DC and AC characterization and validation of the extracted model by designing sample circuits.

The extraction of model includes DC (I-V) characteristics, AC (S-parameter at different bias points), fine optimization of the selected model parameters based on measured device performances and further fine tuning parasitics to match circuit performance at different operating points (bias/input drive).

4.4 Output Filter

The frequency tripler is based on the common source single ended topology for better stability. The low noise HEMT transistor is used for tripler applications. Third harmonic current generation has been done by choosing the bias point for class A operation [6]. Adequate drain termination along with bias is ensured to have voltage distortion instead of current. The input is matched for the first harmonic along with the trap for second harmonic suppression. The output consists of a matching at third harmonic along with a harmonic loaded band pass filter in series with the phase shifter. The concept of optimization technique is used as reported by Johnson et al. [13]. The phase shifter in series with band pass filter provides a parallel resonance of the drain parasitics to maximize voltage at the harmonic frequency. Instead of harmonic filter [14] the concept of harmonic load eases circuit complexity giving clean spectrum. It has been optimized along with the phase shifter to present a short circuit for the even harmonics and an open circuit at the odd harmonic except for the output harmonic. The phase shifter along with harmonic loaded topology of the filter adjusts the impedance to higher values for fundamental, short for second harmonic and transparent to the third harmonic, where $Z=-j[1/\omega C+Z_0\cot(\beta l)]$

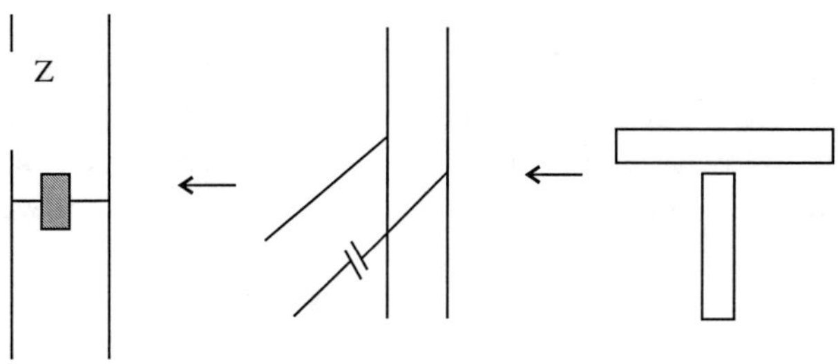

Fig. 4.9: Modelling of stub resonator—concept used in harmonic filter

Band pass filter at $3f_0$ is separately designed, tested and incorporated at the output of the tripler. The filter is based on the overloaded topology to suppress the second harmonic [Fig. 4.9]. Using the harmonic loading concept in the overloaded filter [15], the second harmonic is shorted out and capacitive loading is presented to the first harmonic using electromagnetic simulator [16]. This provides the voltage distortion which is high in odd harmonic current. This filter is incorporated at the output of the multiplier. Instead of harmonic filter the concept of harmonic load eases circuit complexity giving clean spectrum. It has been optimized along with the phase shifter (adding simple line length to adjust the phase) to present a short circuit for the even harmonics and an open circuit at the odd harmonic except for the output harmonic which is passed with minimum loss.

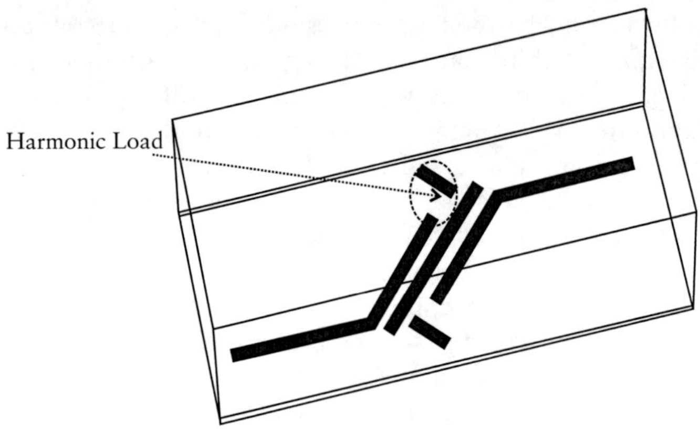

Fig. 4.10: Harmonic-loaded band pass filter

4.4.1 Fabrication and Measurement

Alumina is considered to be rugged and repeatable substrate for RF applications [Chapter 1]. MIC devices are realized on alumina due to better process control and repeatable performance along with lower achievable losses. Development of the substrate and subsequent characterization imposes various challenges due to associated circuit performance which are dependent purely on the dielectric properties [15]. Further surface smoothness finish is also critical parameter due to subsequent thin film operations.

The filter at the output frequency has been tested separately and then further optimized using electro-magnetic tool. The harmonic load is adjusted for suppression of even harmonics. The filter gives insertion loss of better than 1 dB with return loss better than 15 dB. The second harmonic is suppressed by more than 40 dB.

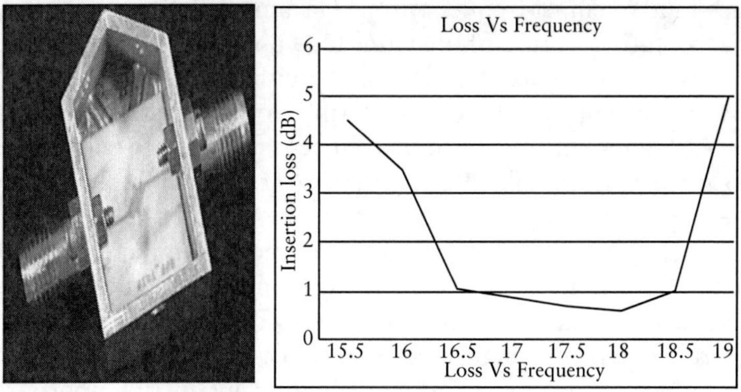

Fig. 4.11: Measurement of harmonic-loaded band pass filter

Harmonic loads are adjusted for suppression of even harmonics and optimum termination for undesired odd harmonics. The designed harmonic loaded filter gives insertion loss of better than 1 dB with return loss better than 15 dB. The second harmonic ($2f_c$) is suppressed by more than 40 dB.

4.5 Circuit Design

The main design consideration in FET-based multipliers is to maximize the transconductance for which device need to be operated at current saturation [16]. The prominent considerations are:

- Bias the device at the pinch off
- Achieving maximum transconductance during the gate biasing with sufficient RF power
- Device to be kept in current saturation.

Methodologies to achieve the above requirements are:

- Conjugate matching of the gate at desired frequency
- Sufficient RF power drive
- Undesired harmonics at drain to be short circuited except desired harmonics.

Gain is given preference over the noise figure in case of multiplier.

The tripler has been designed using the HEMT device which is a low noise device. The input circuit is matched at the fundamental frequency f_0. Harmonic balance analysis incorporating feature of 'dynamic load line' has been performed to have better understanding of the circuit performance. The major criterion for the efficient tripler operation is the choice of bias along with the harmonic current feedback to the device. To get asymmetrical output waveform rich in third harmonic the approach adopted is by biasing the device in class A and presenting short circuits at the even harmonics and open circuits at the odd harmonics, except for the third harmonic which is matched for maximum output power. This voltage distortion approach provides higher odd harmonic current, compared to the other approaches. The major constraint in this approach is placing quarter wavelength transformers for short circuiting the even harmonics. This increases the circuit complexity as impedance at other harmonics also has to be taken into taken into account and should be tuned out. Using the harmonic loading filter and analyzing it using electro magnetic simulator [17], the second harmonic is shorted out and capacitive loading is presented to the first harmonic as described by Camargo [2]. This filter is separately designed, tested and incorporated at the output of the tripler. The electric phase between the tripler and filter is adjusted to have better matching reflected in the output conversion loss.

The phase shifter along with harmonic-loaded topology of the filter adjusts the impedance to higher values for fundamental, shorting for second harmonic and transparent to third harmonic, thus providing efficient tripler. The transmission line acting as a phase shifter in series with band pass filter provides a parallel resonance to the drain parasitics to maximize voltage at the harmonic frequency. The electric phase between the tripler and filter is adjusted to have better matching reflected in the output conversion loss. Phase shifter along with harmonic-loaded concept [section 4.3] of the filter adjusts the impedance to higher values for fundamental, shorting for second harmonic and transparent to third harmonic, thus providing efficient multiplication (Fig. 4.11).

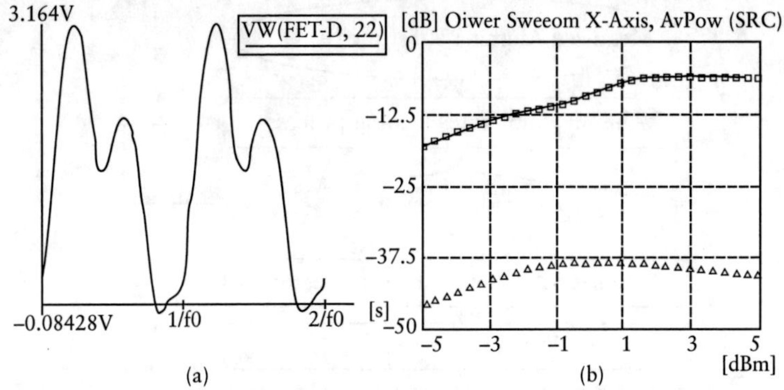

Fig. 4.12: Simulated performance with input (a) drain voltage waveform for +2dBm (b) multiplication gain

4.5.1 Circuit Realization and Measurement

Alumina substrate is chosen for the realization of the tripler circuit. Selection and thickness of substrate is an important consideration and the process steps to have tight dimensional tolerances results in the better performance. Complete process steps are shown in Fig. 4.13.

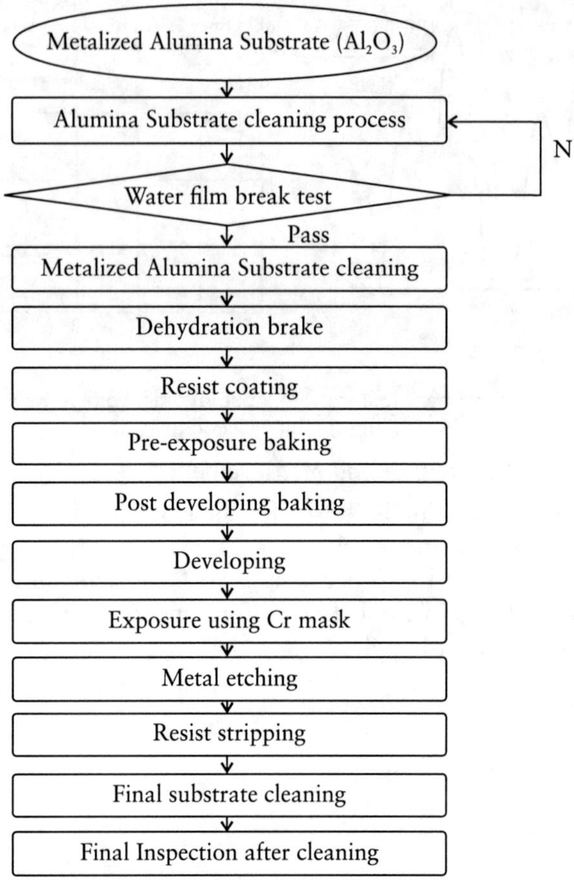

Fig. 4.13: Patterning process steps for circuit realization

This approach is validated by developing two active triplers at two different output frequencies 17 GHz and 13.5 GHz respectively. The class A operation was used by biasing the device at Vds = 1.5 V and Ids = 12 mA.

Simulation study carried out to find out the effect of the band pass filter. The output before BPF and after BPF insertion (Fig. 4.14) shows considerable improvement in the overall response.

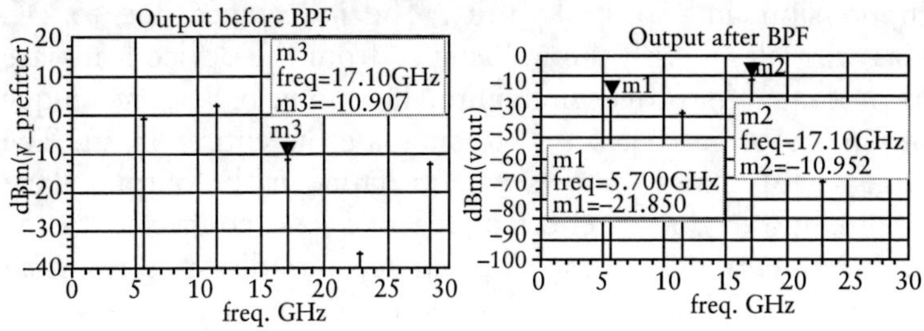

Fig. 4.14: Effect of filter insertion on the multiplier performance

4.5.2 Assembly and Measurement

Tripler circuit along with filter is etched on 10 mil alumina and attached to Kovar plate using conductive epoxy [Fig. 4.15]. The input frequency is 5.7 GHz and output is measured at 17.1 GHz. The measurements are taken for different drain and gate biases. The class A operation having Vds=1.5 V and Ids = 12 mA results in the conversion loss of better than 10 dB in the 3 dB bandwidth of 200 MHz. The input power typically required is +2 dBm for having good efficiency. The suppression at fundamental is better than 15 dB and second harmonic suppression is better than 25 dB. The output power comes out to be –5.5 dBm for an input drive of +2 dBm.

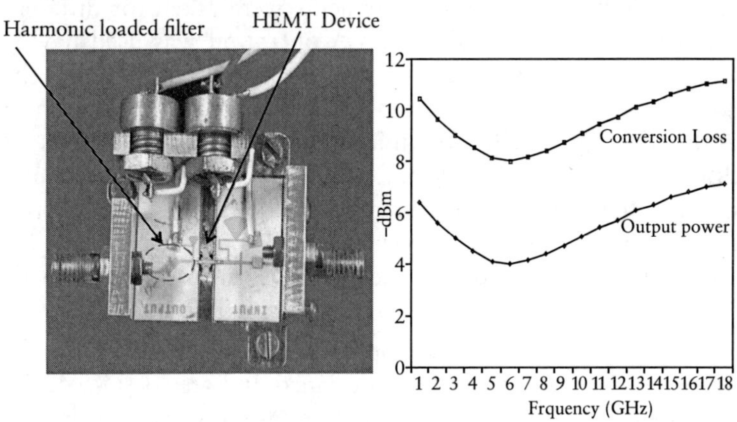

Fig. 4.15: Tripler circuits at Ku band (17.1 GHz)

The full card along with filter is etched on 10 mil alumina and attached to Kovar plate using conductive epoxy. The principal parameters of interests are the conversion loss, harmonics suppression and spectral purity. The inductance due to DC bias wire along with the inductance from the device leads are incorporated for proper simulation. Harmonic balance technique of CAD tool has been used to simulate the entire circuit. The large signal analysis shows the trajectory at the output plane is an ellipsoid due to effect of reactive load connected to the drain. An extremist of the ellipsoid becomes distorted resulting in clipping of drain voltage and drain current [Fig. 4.16].

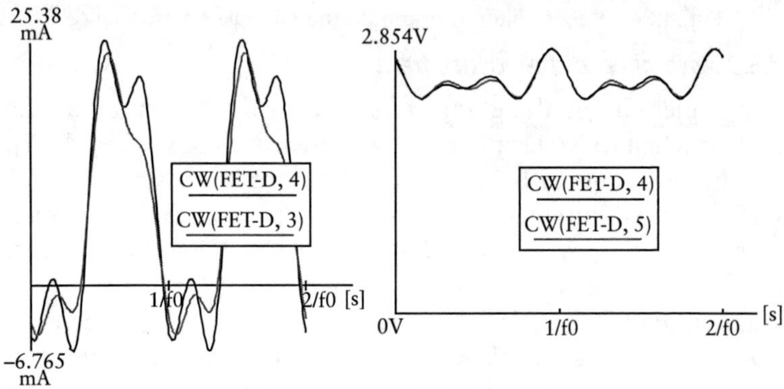

Fig. 4.16: Voltage and current waveform

The input frequency is 5.7 GHz (f_c) and output is measured at 17.1 GHz ($3f_c$). The measurements were taken for different drain and gate biases. The class A operation was used having Vds=1.5 V and Ids = 12 mA. The measured results show the conversion loss of better than 10 dB in the 3 dB bandwidth of 200 MHz [Fig. 4.15]. The input power typically required is +2 dBm for having good efficiency. The suppression at fundamental is better than 15 dB and second harmonic suppression is better than 25 dB. The output power comes out to be –5.5 dBm for an input drive of +2 dBm. The circuit is tuned at 16.95 GHz. The analysis shows the shift due to etching tolerances.

To further validate the approach one more tripler for the output of 13.5 GHz has been designed and tested [Fig. 4.17]. The input frequency is 4.5 GHz (f_c) and output is measured at 13.5 GHz ($3f_c$). The same filter topology is employed and tuned accordingly. Measurements were taken for different drain and gate biases. The class A operation is selected having Vds=1.5 V and Ids = 12 mA. Measured results show the conversion loss of better than 10 dB in the 3 dB bandwidth of 200 MHz [Fig. 4.18]. Input power typically required is +3 dBm for having good efficiency. The suppression at fundamental is better than 15 dB and second harmonic suppression is better than 25 dB. The output power comes out to be –5.5 dBm for an input drive of +3 dBm. This closely relates to the simulated values. The realized tripler topology is shown in Fig. 4.17.

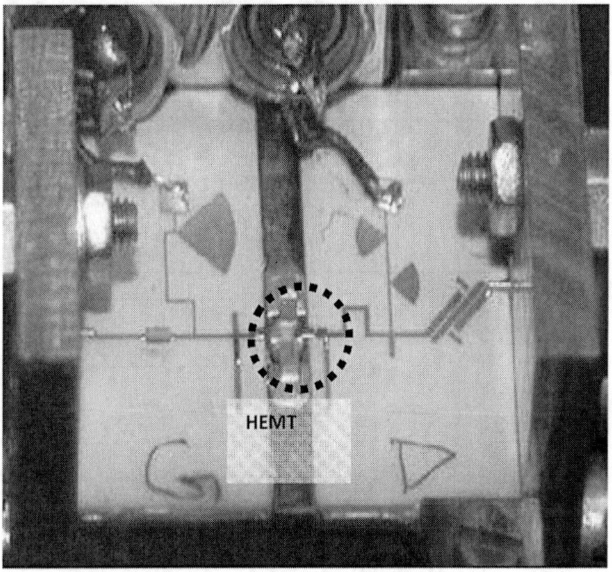

Fig. 4.17: Tripler circuits at Ku band (13.5 GHz)

The measured results show a conversion loss of better than 10 dB in the 3 dB bandwidth of 200 MHz [Fig. 4.18 showing output power at third harmonic with respect to input frequency]. The input drive typically required is +2 dBm for having good efficiency. The rejection at fundamental is better than 20 dB and second harmonic is suppressed better than 30 dB. The tripler provides output power of –5.5 dBm for an input of +2 dBm with the size of around 20 mm × 15 mm. Experiment results are in good agreement with harmonic balance simulated results.

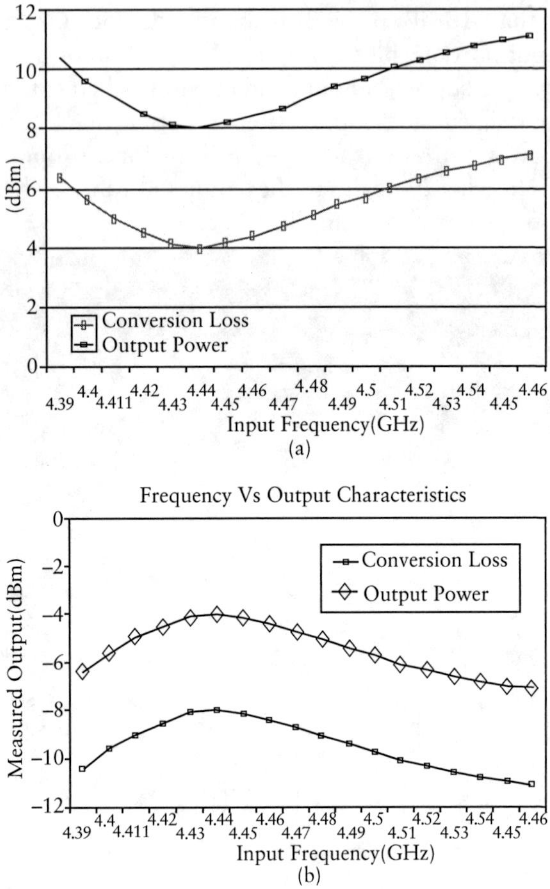

Fig. 4.18: Output power of FET tripler output (a) 17.1 GHz (b) 13.5 GHz

The tripler has been designed using the HEMT device which is a low noise device. The major criteria for the efficient tripler operation are the choice of bias along with the harmonic current feedback to the device. To get asymmetrical output waveform rich in third harmonic presented the approach is by biasing the device in class A and introducing short circuits at the even harmonics and open circuits at the odd harmonics, except for the third harmonic which is matched for maximum output power. Conversion loss of around 8 dB and output power of –4 dB is achieved for the designed triplers. The optimum gate bias point is chosen from the plot as shown in Fig. 4.18. The effect of gate bias on conversion loss is shown in Fig. 4.19.

As shown in Fig. 4.17, drain bias consists of double stack of high-low impedance section for giving better isolation between DC and RF at high frequency end.

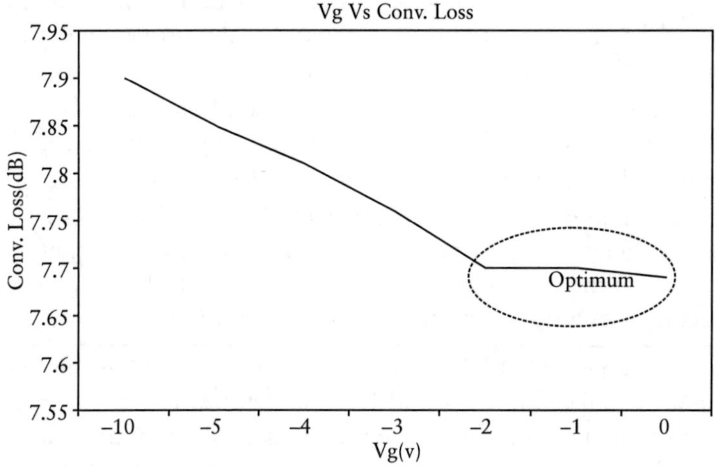

Fig. 4.19: Variation of conversion loss with varying gate bias

Overall, the present approach makes use of simplistic non-linear modelling approach, selective for a particular application to provide efficient non-linear circuits.

Slight deviation from simulated output power is attributed to the leakage of the signal at the bias input. The present approach uses variation of the non-linear parameters and optimization using CAD tools. A stable tripler, having 200 MHz bandwidth is developed with a multiplication gain better than –10 dB at 17.1 GHz. In spite of negative gain presented, the losses comparatively are much lower, as an equivalent diode tripler is reported having –20 dB conversion loss. The poor return loss at the input results in gain reduction and further improvement in the input match along with isolator incorporation can improve overall gain performance. But this feature can be obtained input match and efficiency can be obtained at the expense of the bandwidth. The slight detuning at the lower end is due to under etching. The same technique has been adopted at another frequency, i.e. at

13.5 GHz and measured results show the same performance, at par with simulation.

The presented methodology shows the procedure for the modelling of single gate HEMT which is fine tuned for multiplier operation. The approach uses variation of the non-linear parameters and optimization using CAD tools. The Materka model has been used and parameters for the proper operation of DC and AC have been found out.

Overall, active multiplier design is based on primarily three steps: non-linear device characterization, input match for the gain, and output match for harmonic selection.

The major drawback of tripler based on FET devices is the requirement of high power drive to achieve low sensitivity to device parameter variations. The present approach presents the circuit having optimum performance with a lower drive.

A non-linear HEMT model has been generated using the small signal parameters and DC characterization. First order approximation is assumed which is capable of representing the device non-linearities. A quasi-static approach assuming the time-invariant lumped elements which depend only on the dc bias and amplitude of the terminal dynamic voltages is implemented. The drain-source current is represented by Materka model which is further optimized to have close match between simulated and measured AC and DC data. An accurate model is necessitated to be incorporated in the CAD tools to have a close match between the measured and simulated results. The proposed model also functions as a master linear model that accurately replicates measured microwave S-parameters at arbitrarily chosen bias point within the transistor useful operating I-V range. Active multiplier with an order greater than two is difficult to realize at higher frequencies due to parasitic feedback, dispersion and efficient square wave generation. Also the terminating impedances at the fundamental frequency and harmonics become interrelated. This necessitates an accurate model along with simple circuitry to have an efficient performance and to validate the same tripler is realized at Ku-band.

The tripler (single-ended) assembly was kept simple and compact by incorporating harmonic loads. The output power and DC-to-RF efficiency are likely to become impractically small for large values of multiplication N so mostly upto x 3 multipliers are implemented using transistor. Frequency quadrupler can also be on the similar line with the cascading of two frequency doubler.

Active multipliers can be divided into two groups: balanced and single ended. Balanced multipliers can be used to cancel out either odd or even harmonics, Single-ended designs are more efficient, lower power circuits, but tend to have tighter bandwidth of operation.

References

1. Mike Golio, *RF and Microwave Semiconductor Device Handbook*, CRC Press, 2003.
2. E. Camargo, *Design of FET Frequency Multipliers and Harmonic Oscillators,* Artech House, 1998.
3. D.G. Thomas, GR Branner, Single-ended HEMT RF/microwave frequency doubler design using reflector networks, 38th Microwave Symposium on Circuits and Systems, Proc-2, 1014-1017, 1995.
4. E. Camargo, "Source of Non-Linearity in GaAs Frequency Multipliers," *IEEE Trans MTT-S Digest*, pp. 343-45, 1983.
5. Kamaljeet Singh and R. Ramsubramanian, "Simplified Non-Linear HEMT Modelling and Validation," Mathematical Techniques: Emerging Paradigms for Electronics and IT Industries (MATEIT-2006), Delhi-2006.
6. M.S. Gupta, R.W. Laton, and T.T. Lee, "Performance and Design of Microwave FET Harmonic Generators," *IEEE Trans MTT*, March 1981, pp. 262-63.
7. R. Gilmore, "Concepts in the design of Frequency Multipliers," *Microwave J.*, 1982, pp. 130-36.
8. O.V. Stein and Jim Sherman, "Odd order MESFET Multipliers with Broadband, Efficient, Low Spurious Response," *IEEE Trans MTT-S Digest*, pp. 667-70, 1996.
9. M. Sango. O. Pitzalls, Lennen, Wang and Childs, "A GaAs MESFET Large–signal Circuit Model For Non-linear Analysis," *IEEE MTT-S Digest*, pp. 1053-56, 1988.

10. M.T. Hickson, P. Gardner, and D.K. Paul, "HEMT Model for S-parameter and Noise Parameter Extrapolation," *IEEE Trans MTT-S Digest*, pp. 277-80, 1992.

11. M. Carmen Curras-Francos, "Table Based Non-linear HEMT model Extracted from Time –Domain Large signal Measurement," *IEEE Trans MTT*, May 2005, pp. 1593-99.

12. W. Curtice and E. Ettenburg, "A Non-Linear GaAs FET Model for use in the design of output currents for Power Amplifiers," *IEEE Trans MTT*, Dec. 1985, pp. 1383-93.

13. J.E. Johnson, Branner, and Mima, "Design and Optimization of Large Conversion Gain Active Microwave Frequency Triplers," *IEEE Microwave and Wireless Components Letters*, 2005, pp. 457-59.

14. Alfy Riddle, "High Performance Parallel Coupled Microstrip Filters," *IEEE MTT-S Digest*, 1988, pp. 427-30.

15. Kamaljeet Singh, *Interdigital Capacitor*, Atlantic Publishers and Distributors (P) Ltd., 2020.

16. Steven Maas, *The RF and Microwave Circuit Design Cookbook*.

17. Daniel G. Swansen and W.J.R. Hoefer *Microwave Circuit Modelling Using Electro-magnetic field simulation*, Artech House, 2003.

Chapter 5

Mixer Based Topology

5.1 Introduction

Frequency multiplication happens due to component non-linearities and resultant spectral components are the sum and difference of the input frequencies. Non-linearity is associated with the harmonics so the sum and difference products are created at the multiples of the input frequencies. Mathematically, it can be represented as

$$V_0 = \frac{A_1 A_2}{2}[\cos(\omega_1 - \omega_2)t - \cos(\omega_1 + \omega_2)t]$$

where $V_1 = A_1 \sin\omega_1 t$ and $V_2 = A_2 \sin\omega_2 t$

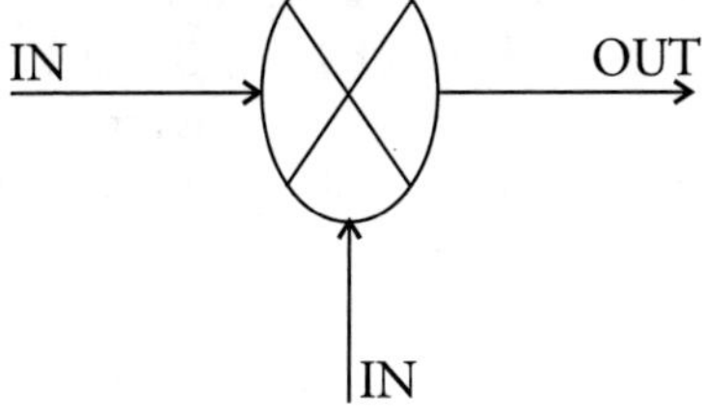

Mixer converts received frequency to the baseband or base band to the transmit frequency using non-linear characteristics of diode or transistor. This phenomena is carried out by injecting RF signal into non-linear device creating harmonics and the desired output to be selected from the harmonics. Mixers are classified as image rejection, balanced, doubly balanced. It is three-port device having two-inputs ports and one-output port.

Mixer converts frequency component into another frequency components for the ease of signal processing. The important parameters in mixer are: conversion loss, insertion loss, return loss, etc. [1].

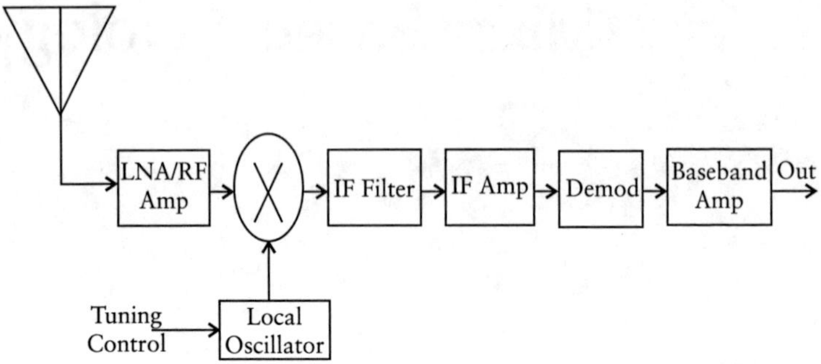

Fig. 5.1: Mixer in heterodyne receiver architecture

5.2 Mixer Basics

Frequency mixer generates signals at the sum and difference of two input signals. Standalone mixer in MMIC forms as well as in MIC can be designed using standard techniques. Mixer output can be represented as

$$f_0 = \pm n f_1 \pm m f_2; \text{ m, n are integers}$$

Selection of mixer topology is based on the applications, noise factor and multiplication order so as to achieve the desired performance without any noise. The circuit can be realized either incorporating diode (preferably Schottky) which provide conversion loss or by incorporating FET providing conversion gain. In superheterodyne topology, mixer is an important component due to conversion of input frequency to IF before demodulation. Mixer performance degradation depends on the local oscillator and signal frequency so that distortion of signal can be minimized. Signal-to-noise ratio of receiver depends upon the mixer performance as reciprocal mixing can raise the noise floor which can limit receiver ability to recover weak signals. The performance of local oscillator and its phase noise is important as mixer is operated at saturated power so amplitude modulations are insignificant.

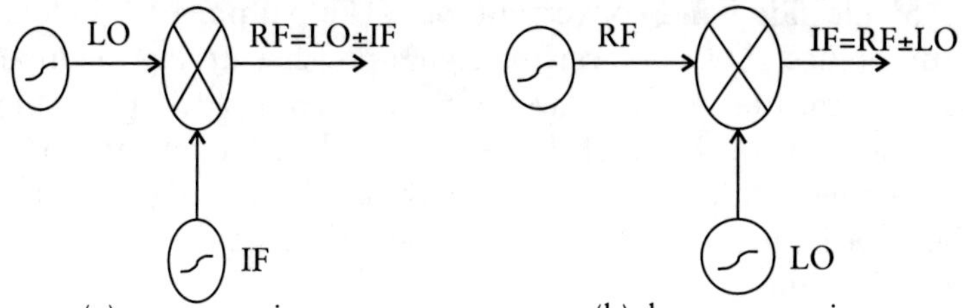

(a) up-conversion (b) down-conversion

Fig. 5.2: Frequency conversion with mixer

The diode-based mixer topology is having the equivalence of the diode model with series parasitic [2]. At low frequency, capacitance is inconsequential and series resistance is dominant. At high frequency both capacitance and resistance are considered. The associated main parameters are dynamic range, conversion loss, bandwidth, noise figure, isolation and VSWR. The main contribution is by the resistance and the conversion loss which can be represented as.

$$L(dB) = 10\log\left(1+\frac{R_s}{R_j}+4\pi^2 f^2 C_j^{\,2} R_s R_j\right) \qquad \ldots (5.1)$$

Mixing the input signals having desired information yields upper and lower sidebands containing the desired information. Two side bands containing identical information namely upper sideband which is sum of the input and LO frequencies and the lower sideband which is the difference between the input and LO frequency is known as IF. Mixer uses Schottky barrier diodes but also utilizes BJT, FET devices. Diode-based mixer are easy for implementation and can be employed in various configurations. Mixer can be four-port device if DC bias for LO operation is employed as normally LO power turns on the diode.

5.3 Mixer Terminology

Schottky barrier diodes are employed for mixing as LO power is sufficient but it results in the conversion loss. Transistors provide conversion gain but are having high noise figure and need external bias. Schottky barrier diodes can be manufactured employing GaAs or silicon. The cut off frequency in GaAs is five times higher than Si but it needs higher forward voltage resulting in higher LO power requirement and higher flicker noise [3].

Single side band conversion loss: The parameter is defined as difference between receiving signal power entering the input port and the out-port IF power of the desired IF sideband exiting the I-port. As only one of the output product are taken so 3-dB SSB conversion loss is achieved (without considering unwanted higher order mixing product) due to series resistance of the diode and mixer mismatch.

The primarily cause of the output noise is due to LO signal phase modulated by noise and measurement of the same is done through SSB phase noise depicting oscillator phase instability. LO employed in FM receiver need not be as low in phase noise as in the case of PSK receiver.

VSWR: It represents the mismatch between the ports. Voltage standing wave (VSWR) and current standing wave ratio is having opposite phase. The VSWR depends on the LO power as the power level (which is usually much higher than RF level) changes the operating bias points resulting in the mismatch at mixer ports and alters the VSWR at varying power levels. So LO port is optimized for the return loss and afterwards RF and IF ports are adjusted for proper matching.

Isolation: It represents unwanted signals and taken between RF and LO port as well as between LO and IF port due to higher LO power level. IF distortion and electromagnetic interferences are caused due to unwanted signals.

Table 5.1: Isolation Vs leakage

Incident power from LO (dBm)	L-R isolation (dB)	Leakage (dB)
10	40	–30
10	30	–20
10	20	–10
10	10	0

As shown in the above table, leakage increase with the reduction in the isolation which can results in lower leakage from receiver to the antenna.

Dynamic range: It is the linear range of the mixer which is defined from the noise floor to the level of 1-dB compression point. The upper point is the increase in the conversion loss which got converted into heat and higher order inter-modulation products. The mixer having low LO power range is having low dynamic range as dynamic range is generally 5 to 10 dB lower than the LO input power.

Inter-modulation products: It is undesired mixer generated output products from the mixer ports and can be classified as single tone or multiple tone. Single input RF signal mixing with the LO products single tone whereas two- or more input RF signals mixing with LO provide multi-tone products. The inter-modulation products levels are effected by input LO and RF power levels and frequencies. Double-balanced mixers due to circuit balance are having much better suppression compared to odd-by-odd products. FOM of inter-modulation is given by intercept point.

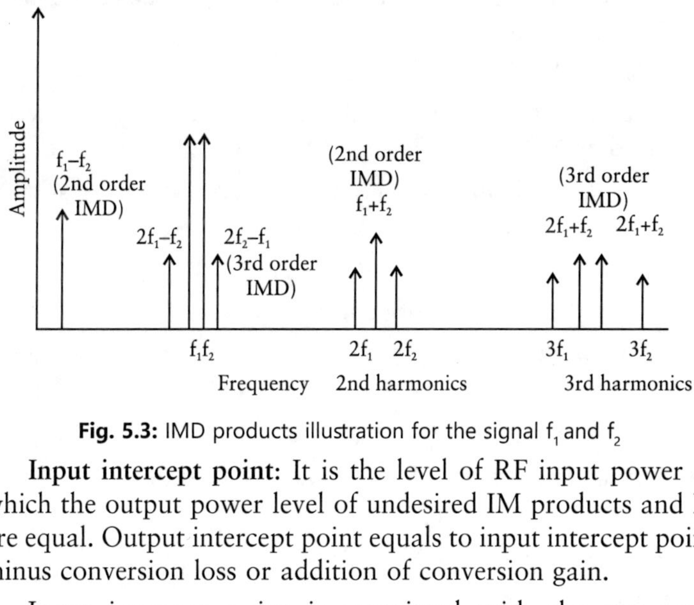

Fig. 5.3: IMD products illustration for the signal f_1 and f_2

Input intercept point: It is the level of RF input power at which the output power level of undesired IM products and IF are equal. Output intercept point equals to input intercept point minus conversion loss or addition of conversion gain.

Input intercept point is associated with the two-tone, three-order inter-modulation products as it is closed to the desired IF products.

Dynamic range, 1-dB compression pints and intercept point are all inter related. Two-tone, three-order inter-modulation products refer to the input RF power and 1-dB change in the power level of each input RF signal causes the power level of each IM product to change by 3-dB [4].

5.4 Mixer Design

Mixer converts frequency component into another frequency components and the same can be represented mathematically as:

The second term give rise to several components such as $2\omega_{RF}$, $2\omega_{LO}$, $\omega_{RF} + \omega_{LO}$, $\omega_{RF}-\omega_{LO}$ from which $\omega_{RF}-\omega_{LO}$ term is taken out termed as IF frequency [5].

Table 5.2: Mixer performance

Variants	Diode qty	Matching	Loss	Isolation
Single ended	1	Poor	Good	Reasonable
Balanced (90^0)	2	Better	Good	Reasonable

Balanced mixer is preferable as it allows cancellation of certain multiplication products.

Large Signal Modelling of Diode

Large LO pump signal leads to time varying conductance expressed as

$$g(t) = \Sigma\, Gn \exp(jn\omega_{LO}\, t) \quad \text{... (4.5)}$$

where $Gn = 1/T\int g(t) \exp(jn\omega_{LO}\, t)$

Similarly, $i_d(t) = \Sigma\, In \exp(j(\omega_{IF} + n\omega_{LO} t))$

$$Cj(t) = \Sigma\, Cn \exp(jn\omega_{LO}\, t)$$

As $i(t) = g(t).v(t)$ ... (4.6)

$$I(t) = \Sigma In\, e^{j(\omega_{IF} + n\omega_{LO} t)} \quad \text{... (4.7)}$$

$$V(t) = \Sigma Vn\, e^{j(\omega_{IF} + n\omega_{LO} t)} \quad \text{... (4.8)}$$

Putting values from (4.5), (4.7), (4.8) into (4.6)

$\Sigma In\; e^{j(\omega_{IF} + n\omega_{LO} t)} = \Sigma\Sigma GmVn\; e^{j(\omega_{IF} + (m+n)\omega_{LO} t)}$

In simple terms, $I_{IF} == \Sigma\Sigma GmVn\; e^{j(\pm n\omega L \pm m\omega_{RF} t)}$

For single balanced case

$i_1 = \Sigma\Sigma GmVn\; e^{j(\pm n\omega L \pm m\omega_{RF} t)} = K$

$i_2 = Ke^{+jm\Pi} = K(-1)^m$

$I_{IF} = i_1 - i_2 = K(1-(-1)^m)$

For even harmonics I_{IF}=0, so even harmonics are suppressed.

Considering AC diode circuit consisting of three components at the frequencies ω_{RF}, ω_{IF} and ω_{IM}.

$$I(t) = I_{RF}\cos\omega_{RF}t + I_{IF}\cos\omega_{IF}t + I_{IM}\cos\omega_{IM}t \quad \dots (4.9)$$

Simplification at the diode ports leads to

$$I_{RF}=G_0(V_{RF}-I_{RF}.Rg)+G_1(-I_{IF}R_{IF})+G_2(-I_{IM}.Rg)$$

$$I_{IF}=G_1(V_{RF}-I_{RF}.Rg)+G_0(-I_{IF}R_{IF})+G_1(-I_{IM}.Rg)$$

$$I_{RF}=G_2(V_{RF}-I_{RF}.Rg)+G_1(-I_{IF}R_{IF})+G_0(-I_{IM}.Rg)$$

where Rg is the source resistance, R_{IF} is the IF port termination as well as image termination.

Taking only IF and RF frequency terms from Eq. (4.9) gives

$$I_{IF}= G_0V_{IF} + G_1 V_{RF}$$

$$I_{IF}= G_1V_{IF} + G_0 V_{RF}$$

For simultaneously matched RF/IF ports leads to

$$Gc=2.[(1+\sqrt{1}-(2.G_1^2/G_0(G_0+G_2))\ /(1-\sqrt{1}-(2.G_1^2/G_0(G_0+G_2))]$$

Calculation of Gc and plotting with V_{LO} ideally to be 3dB. An additional losses due to coupler and series resistance makes it near to 7dB. Also undepleted region of the device and various contact package resistance will also appear in series with non-linear junction.

5.4.1 Single-ended Balanced Mixer

Mixer operation depends on proper control of the impedances at each of mixer terminals along with each of the frequencies of interest. Generation of frequencies termed as modulation products, exists according to the heterodyne principle by which mixer operates [1]. Balanced mixer can be designed either using 90° hybrid coupler or using 180° hybrid coupler. The advantage of 180° coupler is the perfect isolation between RF-LO over wide frequency range and rejection of all even-order inter-modulation products [6]. Present design uses branchline coupler, Schottky diodes and low pass filter. Biasing of the diodes is taken care by LO amplitude so no coupling/biasing capacitor is needed. Two anti-parallel diodes are incorporated in the output ports of the branch line coupler with necessary DC return paths. Quarter wavelength open circuited stubs at fundamental and harmonics of the local oscillator frequencies are added at the IF port of the mixer. It is observed that terminating the second and third local oscillator harmonics and associated mixing frequencies in a short circuit results in reduction of harmonic distortion. The circuit was optimized using non-linear harmonic balance tool. A 90° line at input RF frequency is added to one of the output port of the coupler so as to obtain good VSWR at the RF and LO ports and also to get good isolation between them. The configuration of the mixer is shown in Fig. 5.4.

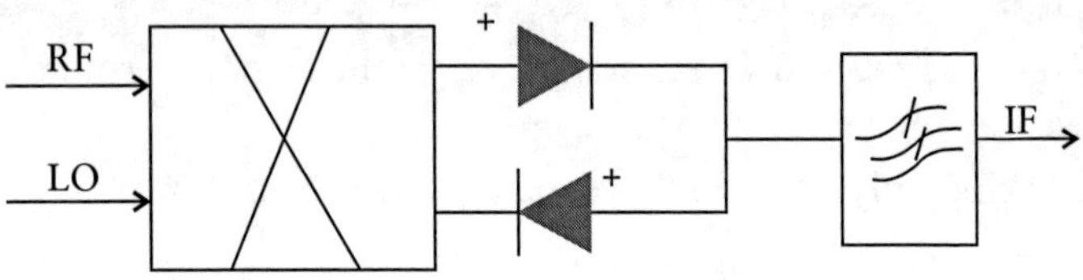

Fig. 5.4: Balanced mixer circuit (90^0 hybrid)

5.4.2 Broadband Coupler

Realization of a compact, broadband mixer, baluns/hybrid has a greater role in overall performance of the overall mixer performance. At higher frequencies increased even mode capacitances reduces port to port isolation [2]. Multiplexing function in the mixer is implemented using an RF-coupler or hybrid junction to provide combining as well as isolation between the two inputs [Fig. 5.3]. Conventional 3-dB hybrid gives equal coupling at just the desired frequency. A broadband hybrid is designed for this purpose to achieve bandwidth of 30% and return loss as well as isolation of better than 20 dB. Equation for designing a hybrid coupler is

$$1/Z_1^2 - 1/Z_2^2 = 1/Z_0^2 \qquad \text{... (4.10)}$$

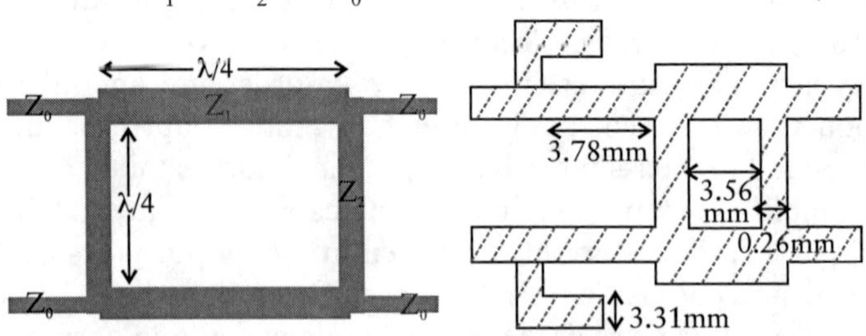

Fig. 5.5: Standard and compensated 3-dB hybrid coupler

where Z_1 and Z_2 are the impedances of series and shunt arms and Z_0 is the characteristics impedance of the line. Putting Z_2=Z_0 the equation leads to Z_1=Zo/$\sqrt{_2}$, which is the case for a 3-dB conventional coupler.

Broad banding is achieved by addition of double quadrature transformers, causing the central match to be scarified for two match points either side of the design centre frequency which can be controlled. The eigen vectors that arises will be identical to that of normal hybrid. These eigen vectors is used to calculate eigen admittances from which eigen reflections are formed. This approach gives a poor match at the centre frequency. This can easily overcome having double quarter wave sections with that of half wavelength matching sections at each end [Fig. 5.6].

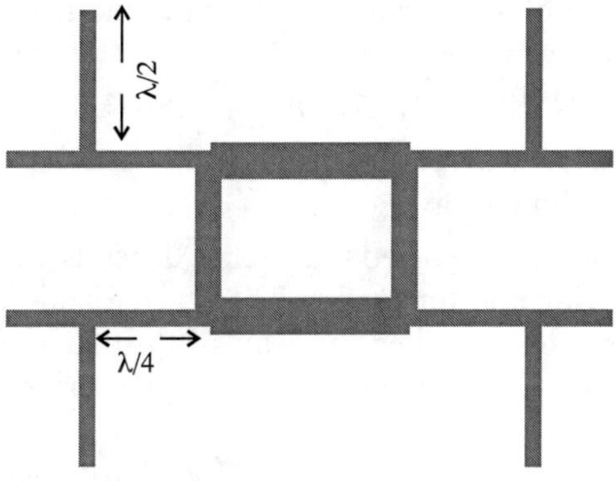

Fig. 5.6: Broadband 3-dB coupler

The reflection coefficient at the centre frequency is

$S_{11}(f_0) = [(1/C^2-1)(Y^2-1/Y^2)]/4+(1/(C^2-1)(Y+1/Y)^2$

C is the coupling coefficient and Y is the eigen admittance which is independent of the frequency. This implies that if the junction can be matched at number of frequencies by an appropriate matching networks, then coupling must be same at all frequencies and depends only on the admittance. Further widening of bandwidth can be achieved by adding half length open stub [7]. The stub length is chosen around 180° for broad band operations. This methodology reduces the undesired coupling in the overall assembly along with maintaining the compactness of the structure. Electromagnetic software has been used to optimize the performance giving a bandwidth better than 25% with better than 20 dB return loss. A branch line coupler is realized on the above mentioned principle to operate in 12 GHz-16 GHz frequency band and the design was analyzed and optimized using electromagnetic simulator. The proximity coupling effects and the discontinuity effects were considered during the analysis. A 90° line at input RF frequency is added to one of the output port of the coupler so as to obtain good VSWR at the RF and LO ports and also to get good isolation between them [8].

In order to achieve overall compactness of the mixer, intermediate frequency (IF) filter is implemented with planar microstrip elements, having low pass filter characteristics.

The design of a single air bridge spiral inductor is based in the formulae given below:

$$L(nH)=0.0008N^2S\,[\ln(S/b)+0.726+0.178(b/S)+(1/8)\,(b/S)^2] \quad \ldots (4.11)$$

$S=Si + Np$, where p=pitch, S=mean side of the spiral, Si=innermost dimension

The value of the realized inductance depends on the number of turns, spacing and line width. The spacing and line width of the transmission line is kept at 5 mils [9] to have more nearer to practical data. Five poles filter employing printed inductors and discrete capacitor has been designed with cut-off of around 1GHz, having chebyshev response for sharper cut-off slope [10]. Insertion loss method is used and the values are obtained for the given specifications. The values achieved for the given specifications are: L1=8.2 nH, and L2=26.5nH which are realized using 2.75 and four turns. The spacing and width is kept around 6 mils after optimization. Effect of capacitance variation on cut-off has been studied and it was seen that the cut-off becomes less than 1 GHz by inserting C1=4.7 pF. Figure 5.7 shows the equivalent of the five pole low pass filter topology.

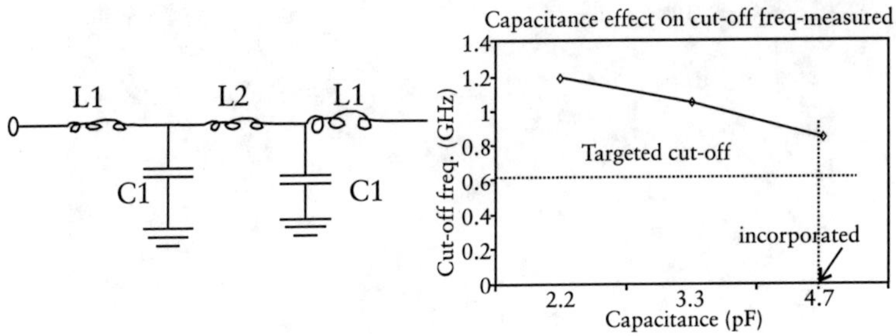

Fig. 5.8: Schematic of five pole LPF and effect of capacitance on cut-off

The printed inductors are placed on the 10 mil alumina substrate having the permittivity of 9.8. The capacitor is mounted in shunt along with the other port grounded through wrap around [Fig. 5.9]. The effect of capacitor Q on overall circuit quality factor is neglected due to greater inherent Q value of chip capacitor. The series resonant frequency is found out from the Smith chart which comes out around 600 MHz. The calculated value of Q comes out around 30.

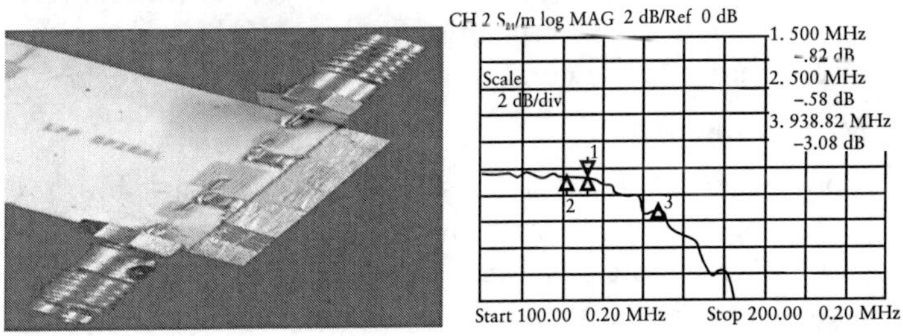

Fig. 5.9: Planar lumped realization of the filter and measured result

The spiral inductors are characterized by impedance measurements. The typical ranges of spiral inductors values realized are 8 nH to 26 nH.

The typical realized values of printed spiral inductors are 8 nH and 2.6 nH which are achieved by the selection of turns.

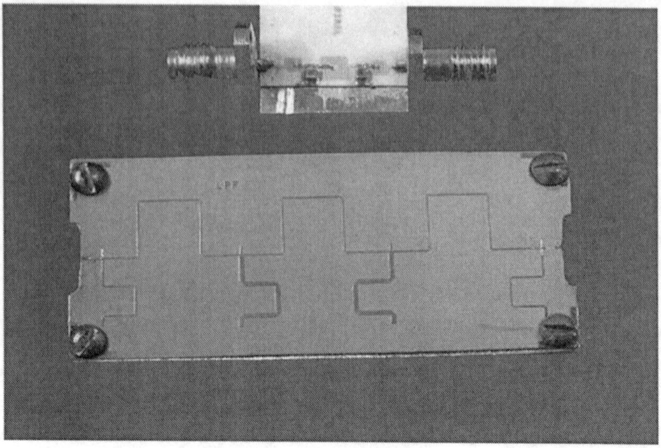

Fig. 5.10: Comparison with distributed topology

The circuit is compared with the realized distributed topology and one-fifth size reduction has been achieved as shown in Fig. 5.10. Various simulation techniques such as SPICE, harmonic balance, MOM, spectral domain are employed for the analysis of the complete assembly. In the present case, MOM and harmonic balance techniques are employed.

5.5 Measurements

The mixer circuit was designed and realized for a RF frequency of 14.5 GHz with an IF frequency of ~600 MHz. This compact circuit has been realized on a 10 mils alumina substrate using thin film technology along with filter [Fig. 5.10].

The measurement of the mixer was done using an automated software taking cable losses at all frequencies into account [Annexure 4]. Different measurement like RF frequency versus conversion loss, IF versus conversion loss and LO power versus conversion loss were carried our using the software. The conversion loss of 8.0 dB over the desired frequency range has been measured.

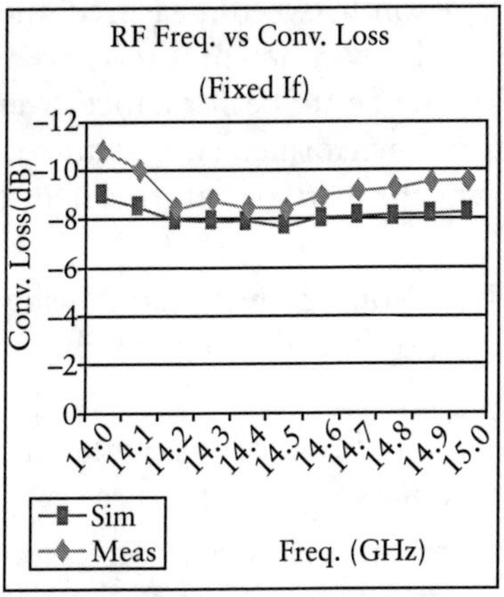

Fig. 5.12: Frequency vs. conv. loss

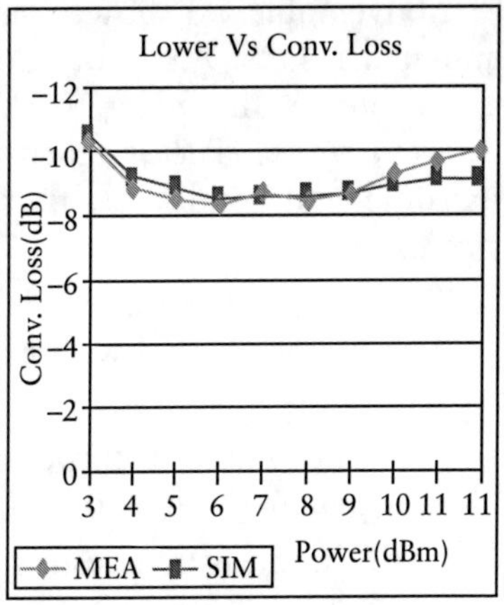

Fig. 5.13: LO power vs. conv. Loss

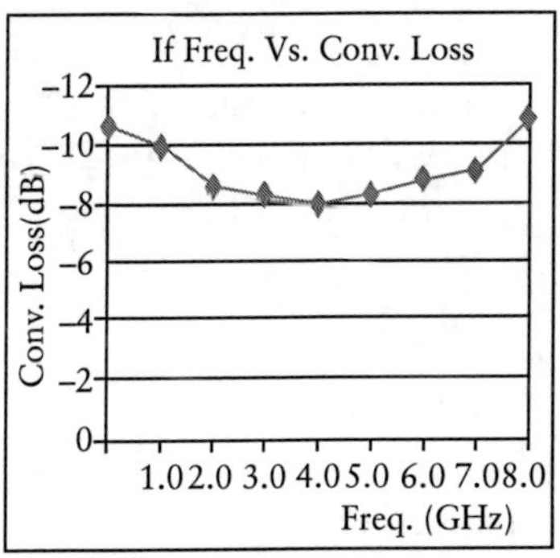

Fig. 5.14: IF freq. vs. conv. loss

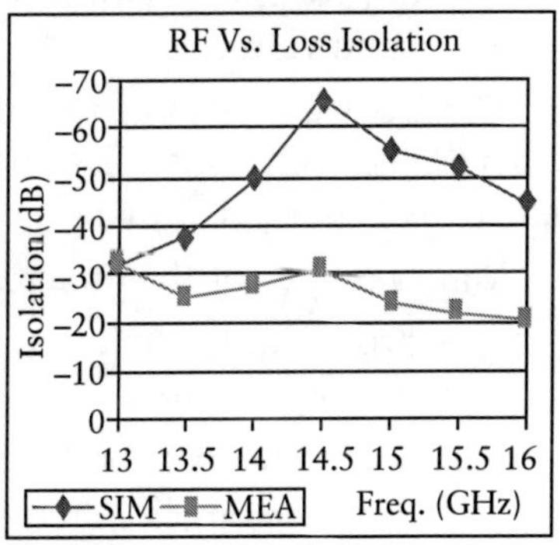

Fig. 5.15: RF – LO isolation

Figures 5.12-5.15 give the comparison of different measured parameters with the simulated results. The measured values are isolation of better than 20 dB and IM suppression of better than 35 dB.

Table 5.3: IM/Harmonic rejections

FREQ (GHz)	O/P SPECTRUM (dB)	
	simulated	measured
13.308	–70	–75
13.903	–36	–42
14.496	–46	–50
27.806	–55	–50

Above Table 5.3 shows comparison between measured and simulated results. Also hot and cold environmental tests were carried out on the mixer to characterize its performance over –30°C to +60°C. The designed circuit performed successfully in environmental conditions, thus will be part of future satellite systems.

The design of the mixer is based on Schottky diode using non-linear diode model with extraction methodology as shown in Chapter 3:

- The mixer is incorporated with broadband coupler and planar IF filter for compactness of the structure.
- The measured results of the mixer is the conversion loss of 8 dB, harmonic suppression of better than 40 dB, RF-LO isolation of better than 20 dB in the 20% bandwidth.

The LO signal is much higher than the RF signal so as to achieve better inter-modulation suppression which can 20 dB higher for better performance.

Main features of the mixer are:

- Broadband coupler (12-16 GHz)
- RF-LO isolation better than 20dB
- DC return paths adjusted for parasitic cancellation
- Pair of open circuited stubs for harmonics
- Optimization for minimal harmonic impedances
- Planar IF filter having cut off around 1GHz.

5.6 Multiplier using Mixer

Present era demands and pushes systems to be designed and operated at higher frequencies. RF systems particularly receivers and transmitters are critical for communication systems. Major components in regenerative transponders are front end followed by down-converter stages. Generating stable RF frequency is the main challenge in these systems. As higher frequency source such as DRO, etc. are stable but are having higher phase noise; so for space-borne applications the better approach is to employ TCXO and frequency multiplication. Frequency multipliers are mostly diode or active device based which are tuning intensive and are not efficient enough to generate higher frequency components. PLL-based approach is also having higher phase noise [11-12]. Step Recovery Diode-based approach is device dependent and repeatability issues along with low throughput are the major concerns. This section presents different but efficient approach of generating high order harmonics which is a mixer-based frequency multiplication technique.

The most crucial stage in transmitter topology is frequency multiplier which converts the fundamental frequency produced by TCXO to a higher frequency. In existing transmitter, Step Recovery Diode (SRD) is employed for frequency multiplication of higher order (n>3). Step recovery diode (SRD) is semiconductor PN junction diode having the ability to generate extremely short pulses from an original signal that are very rich in harmonics.

Mixer-based frequency multiplication technique is designed, simulated and tested and can be considered viable replacement of step recovery diode (SRD) based topology. Proposed approach is designed and simulated using system simulation software, and hardware is developed for achieving response at C-band frequency.

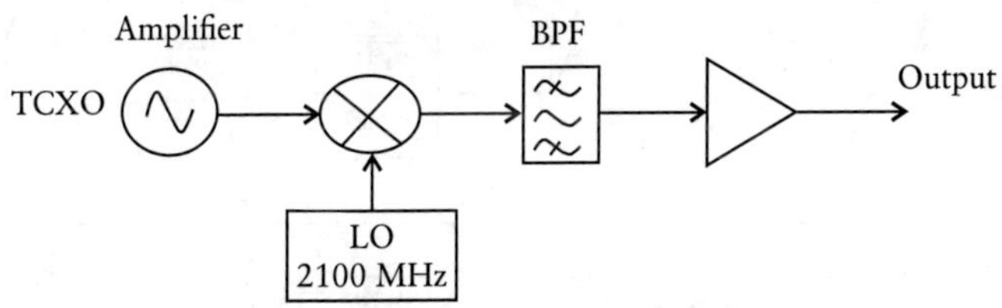

Fig. 5.16: Multiplier schematic using mixer

5.6.1 Simulation Study

The simulation and optimization of the mixer is difficult due to various limitations such as inadequacy of non-linear parameters, complex computational and needs proper simulation set-up to overcome this limitation. Simulation study was carried out to evaluate the frequency response of output signal. SRD-based multiplier and the proposed mixer based frequency multiplier shows the response as shown in Fig. 5.17.

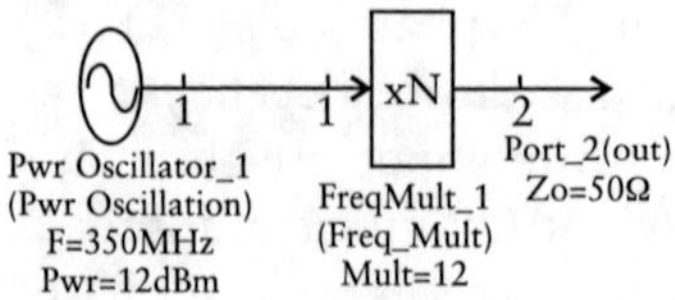

Fig. 5.17: SRD-based multiplier model

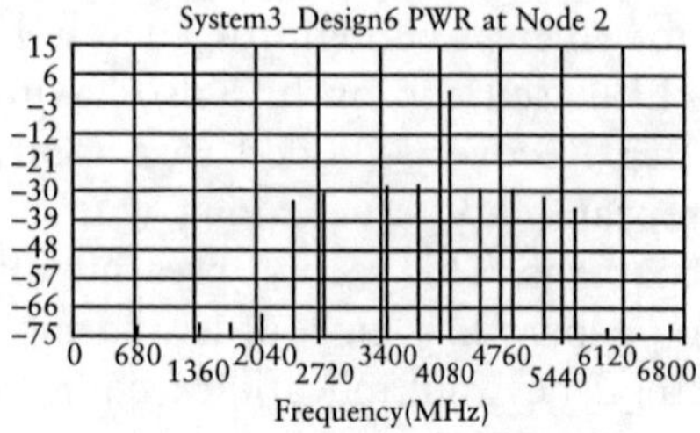

Fig. 5.18: Frequency response

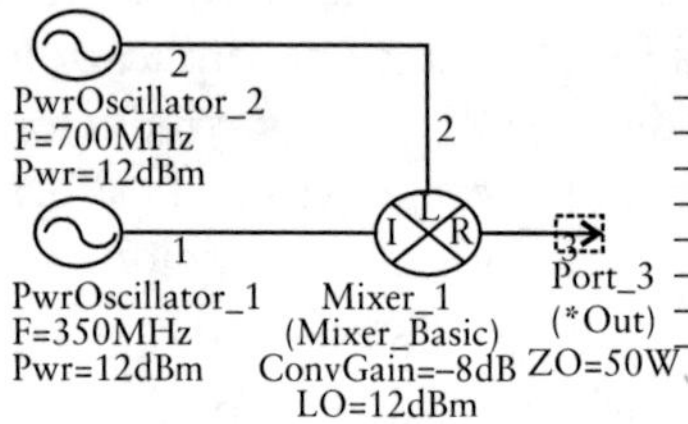

Fig. 5.19: Mixer-based multiplier model

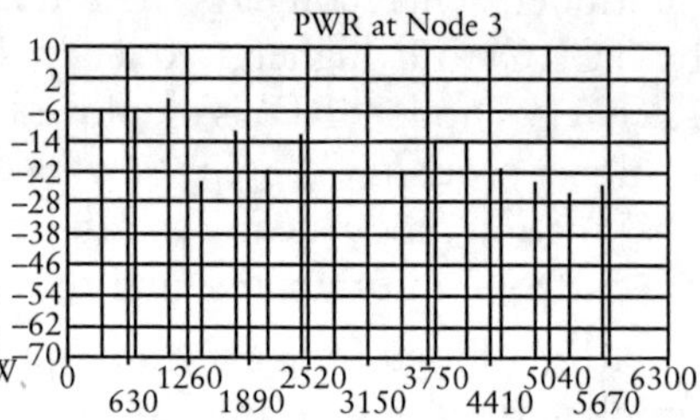

Fig. 5.20: Frequency response

Transmitter model simulated setup with two types of output indicated as low power and high power output.

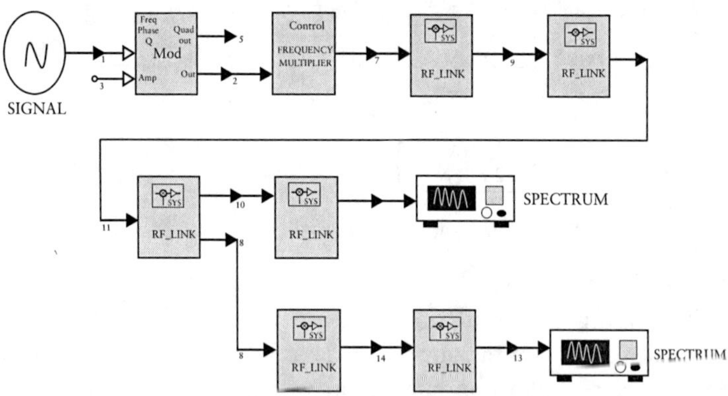

Fig. 5.21: Complete simulation set-up

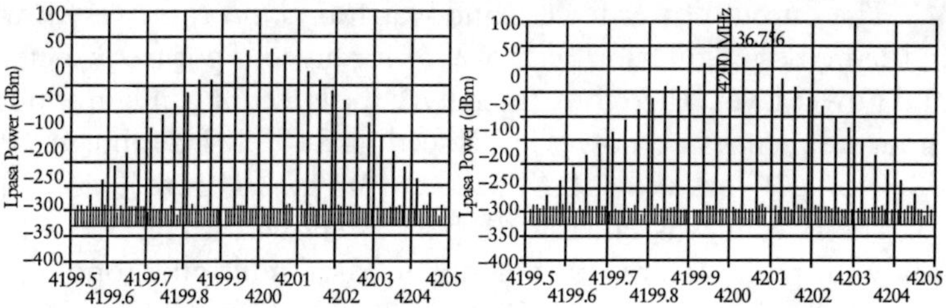

(a) Output at port-1 (b) Output at port-2

Fig. 5.22: Antenna port output

5.6.2 Experimental Study

Experimental-based analysis carried out to validate mixer-based frequency multiplication approach for which a connectorized version mixer was used and frequency response was observed by connecting LO port to output port of first Signal Generator and IF port to the output port of second Signal Generator using calibrated RF cables.

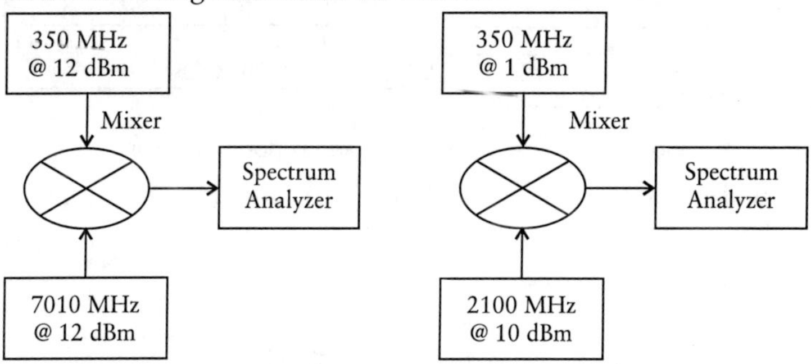

(a) (b)

Fig. 5.23: Experimental set-up

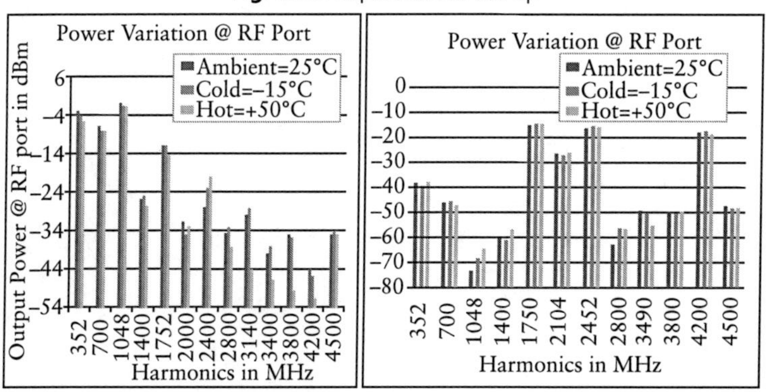

Fig. 5.24: Power variation at different temperature (a) first configuration (b) second configuration as shown in 5.23 (b)

The output power for C-band was found out to be very low in the first configuration (Fig. 5.23a) for given input frequencies and power level. Therefore, to analyze the effect of configuration, input frequency and power levels feeding 2100 MHz input @ 10 dBm for LO Port and 350 MHz @ 1 dBm for RF port (Fig. 5.23 b). The output was taken at IF port as shown in Fig. 5.24. To study the temperature effect on mixer-based multiplier topology, hot (+50°C) and cold (–15°C) cycles are performed. Significant improvement was observed for this configuration. Further effect of power variation at RF and IF port with input power at LO and RF is shown in Fig. 5.25.

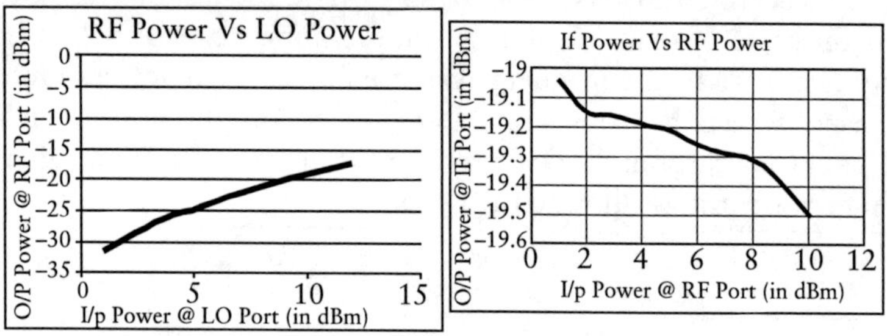

Fig. 5.25: Power variation at various ports of mixer

FET-based mixer topology (Fig. 5.26) can also be implemented but having complexity and needs extensive optimization [13-15].

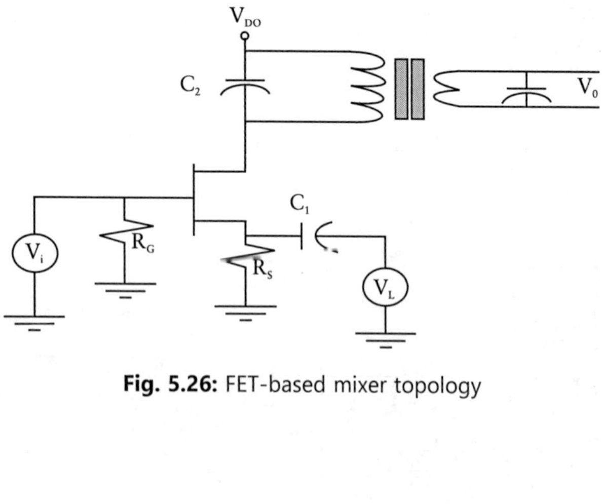

Fig. 5.26: FET-based mixer topology

This approach for frequency multiplication can be implemented easily and needs minimum modification in the existing topologies. Further wide range of mixer models is available for simulation. This methodology eliminates various limitations associated with SRD such as temperature dependency and batch-to-batch variation. The realization time for implementing mixer-based multiplier is lesser due to avoidance and dependence on multiple tuning. This approach aims at yielding a reliable and reproducible package which in turn leads to standardization for enhancing throughput in production mode.

Mixer Vs FET/Diode

Mixer as well as diode/FET-based multiplier rely on the non-linear behaviour of semiconductors but frequency mixer based on subtraction or addition of the two input signals whereas multiplier (diode or FET based) is based on the single frequency.

Mixer is a three-port device in which the frequency translation takes place either in single-sideband conversion or double-sideband conversion having output filter to block unwanted signals.

Higher frequency translations are carried out by frequency multipliers whereas dividers are employed for lower frequency. Compared to FET, diodes are better non-linear devices due to low noise figure as they are operated under local drive in a passive mode.

Conversion loss of a frequency mixer is higher than frequency multipliers and depends on the performance of the LO power. Major LO parameters such as phase noise, spurious and harmonic content impact the mixer performance.

Lumped Equivalence of the Discontinuities

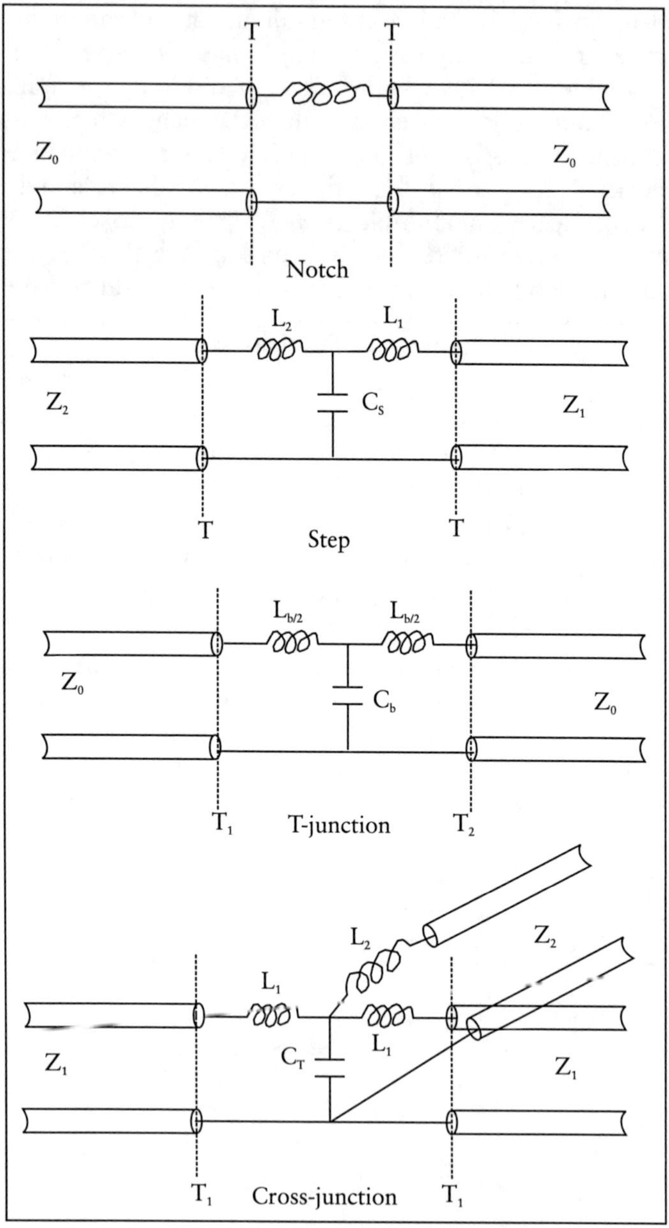

References

1. D.M. Pozar, *Microwave Engineering*, John Wiley & Sons, 2010.
2. Steven Maas, *The RF and Microwave Circuit Design Cookbook*, Artech House 1993.
3. WJ Tech note, Local oscillator for phase noise and its effect on receiver performance, 1981.
4. Inder Bahl, Prakash Bhartia, *Microwave Solid State circuit Design*, Second Edition, Wiley Interscience.
5. Mike Golio, *RF and Microwave Semiconductor Device Handbook*, CRC Press, 2003.
6. Wright, Judah, "Very Broadband flat coupling hybrid ring", Electronic Letters, 1987, pp. 47-49.
7. Asworth, "Design equation to realize a broadband hybrid ring or a two branch guide coupler of any coupling coefficient", Electronic Letters, 1988, pp. 1276-77.
8. Boeck, Behren and Tempel, "Planar microwave mixers for wireless and sensor applications", *Proc. IMOC 2001 IEEE-MTT-S*, pp. 25-36.
9. Har Dayal and Quang Le, "Printed Inductors on Alumina substrate", *IEEE Microwave Magazine*, pp. 78-86, June 2001.
10. Inder Bahl, *Lumped elements for RF/Microwave Circuits,* Artech House, London 2003.
11. D.G. Thomas, G.R. Branner, B.P. Kumar, *Microwave and RF Multipliers*, Wiley, 2005.
12. Manfred Thumm and Werner Wiesbeck, *Frequency Multipliers.*
13. Alexander Chenakin, *Frequency Synthesizers: Concept to Product*, Artech House 2011.
14. Jack Browne, *Deciding Between Mixers and Multipliers/Dividers, Microwaves & RF*, December 2016, pp. 62-65.
15. Robert C. Dixon, *Radio Receiver Design*, Taylor and Francis.

Chapter 6

System Engineering Concept

6.1 Introduction

RF sub-systems are the critical components in the spacecraft as they are employed from the link establishment to payload operations. The main challenge is to have reproducible, reliable and robust system at various frequency bands which needs standardization and understanding to achieve the desired performance. As the packages contain lot of active circuits for which there is batch-to-batch and performance variations which necessitates the effort to tune the performance overcoming the parasitic involved in the actual performance [1]. Systems for space missions consist of transmitters, receivers, associated antenna and feeding systems. Transponders are basically a system which receives a signal, amplifies it, changes its frequency, and re-transmit it and the main components in the telemetry and telecommand transponders are: transmitter, receivers, TCXO, main antenna, null filling antenna, uplink coupler, downlink coupler, preselect filter, global horn, modulator for Ku-band beacon.

The major challenge in RF sub-system is to meet the specifications at the desired frequency covering the requisite bandwidth. This stems out of the fact that frequency vary for each applications and accordingly the tuning of the components and matching plays an important role in meeting the desired specifications. Further fabrication aspect also to be critically examined as specific process such as Step Recovery Diode (SRD) mounting, which is used for multipliers chain, Temperature Compensated Local Oscillator (TCXO) needs careful mounting as they are dedicated for the specific applications [2]. Tuneable and selectable components (mostly passive) need to be carefully tuned/selected so as to achieve the optimum performance in terms of gain, amplitude flatness, modulation index, harmonic and spurious suppression, etc. The deviation in the performance necessitates systematic approach to diagnose and rectifies the problems. Further, active devices are also prone to frequent failure due to improper handling and biasing. Some of the failures encountered in RF packages are listed below:

- Device failure with initial biasing: due to high voltage/ shorting phenomena
- Batch to batch variations in active devices such FETs, BJTs, SRD: matching/tuning to rectify
- Grounding problem: results in oscillation
- Varied behaviour in passive components: component replacement and failure analysis
- Failures in thermal cycling, vibration tests, shock: non-destructive diagnosis and replacement.

Apart from the above, at the higher frequency external effects and lead parasitic, mutual coupling, radiation and associated losses increases considerably in the RF packages. Also proper heat dissipation in power amplifiers circuits to be taken care for avoiding thermal runaway. The most critical system is the multiplier chain and a small deviation needs to be rectified so as to avoid deterioration in the overall system performance.

The generic communication system consisting of receiver and transmitter is shown in Fig. 6.1 and Fig. 6.2 respectively.

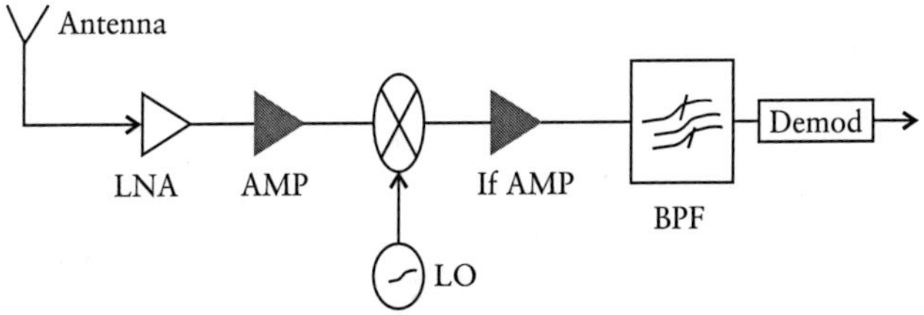

Fig. 6.1: Super-hetrodyne receiver schematic

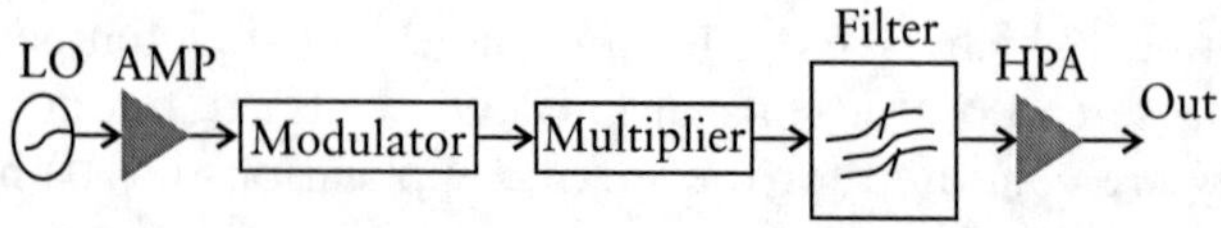

Fig. 6.2: Transmitter schematic

Basic elements in the communication system are: Diplexer, Low noise amplifier, RF amplifier, High frequency Oscillator Mixer and Multipliers. Design of LNA, RF amplifier, DR-based oscillator and system level simulation is detailed in the sections. As single antenna can be employed for transreceiver applications which needs diplexer to be incorporated providing better isolation. This chapter covers all the main topologies and its realization aspects in the microstrip configuration along with detailing of simulation set-up and realization aspects.

6.2 Planar Diplexer

Modern communication systems needs novel filters, diplexers and multiplexers to meet the electrical performances, size and manufacturing cost which is difficult to achieve using the classical topologies. Diplexers are widely used in communication systems for reducing mass and volume of the required hardware. They enable use of the same antenna for different frequency bands, resulting in more compact systems. Their function is to separate and routing the different bands of a signal into different ports. Alternatively, a diplexer combines two different signals with different spectral components into one common port[3]. RF front end of the typically transreceiver architecture is shown in Fig. 6.3.

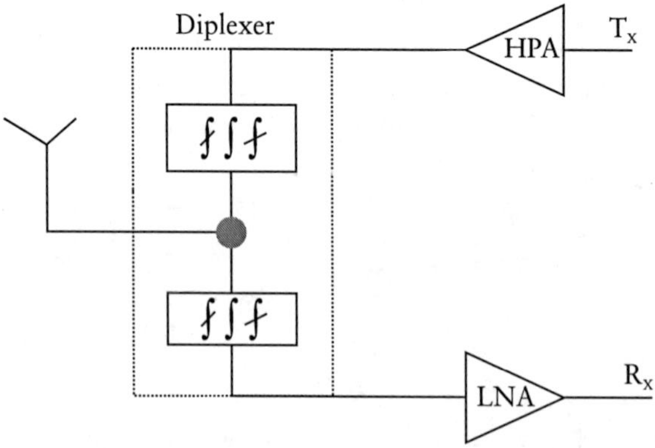

Fig. 6.3: Basic components of transreceiver front-end

The transmitter and receiver operates in different frequency bands and are duplexed to the antenna by the transmit-receive diplexer. The diplexer consists of power divider and two channel filters that can have very stringent specifications. The transmit filter must reject out-of-band noise generated in the power amplifier, so it must have a high level of stop band attenuation, particularly in the receive band. It should have a low pass band insertion loss in order to maximize the DC to RF efficiency of the power amplifier. On the other hand, the receive filter should have high attenuation in the transmit band in order to protect the front-end low noise amplifier (LNA) from the large transmit signals which would otherwise saturate the receiver or can produce high levels of distortion. The major important parameter in the diplexer design is the isolation between receive and transmit channel[4-5]. The main objective of the diplexer unit is to have high rejection for the transmit frequencies in the receive band while preserving the receive frequencies. The same is true for the transmit frequencies, to be passed with low loss, eliminating the receive frequencies [Fig. 6.4].

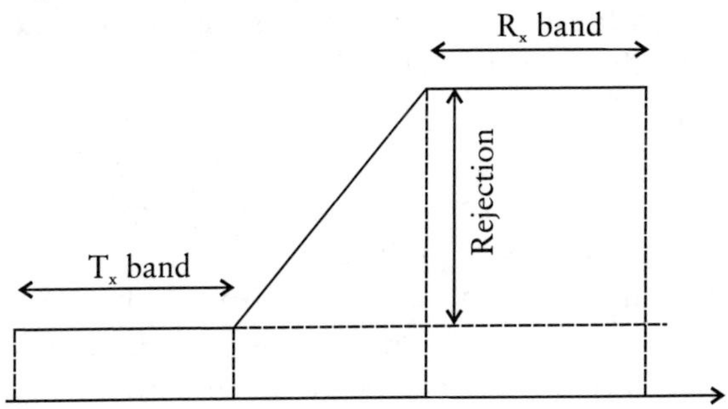

Fig. 6.4: Band specifications

Waveguide filters are the alternative solution and used till now, due to very good characteristics in terms of insertion loss at the expense of the weight, volume and cost. In near future, waveguides in the aerospace equipment will be replaced by planar structures to a large extent. So a planar solution would be better solution to achieve the desired specifications of insertion loss, return loss and rejection. Parallel coupled line (PCL) transmission line can be used to construct many types of filters. PCL approach using the standard classical techniques is not suitable for the diplexer application as it do not allow for reaching the desired specification of narrow bandwidth. Alternatively, multi-section bandpass coupled line filter can be employed having narrower bandwidth, as required for the diplexer applications but limitation of this approach yields large size and uncontrollable transmission zeros resulting in poor selectivity [1]. Hairpin topology is considered to be another approach for placing the TZs, but desired rejection is still not achieved. A modified hairpin topology is proposed [6, 7], using the cross-coupling effect but it increases the circuit complexity considerably and needs tight fabrication tolerances.

Present topology is based on the design and development of the microstrip diplexer using the DBR approach [7]. To achieve the desired specifications, two filters at Rx and Tx frequency bands are optimized. Notch effect on the filter performance is carried out and further implemented in the receive filter. To achieve the desired specifications, notch along with concept of oversized stub can be implemented to have better rejection characteristics in both receive-transmit band. Further suitable T-combiner to be designed for integration of the Rx and Tx filters for the diplexer arrangement.

The specified criteria of insertion loss and flatness have to be better than 3-dB and 1-dB, respectively. The isolation requirements between the two bands are more than 20 dB. The main challenge in this approach is to get the better rejection while restricting the number of resonators to be used so as to reduce both the level of insertion loss and the structure size. The required specification can be met with DBR approach using the oversized stub, which allows increasing the rejection level without altering the TZ frequency [8].

6.2.1 Design Methodology

Diplexer design starts with the design of channel filters (bandpass filter), which must achieve the required pass band performances before they are connected into the power divider. Basically, two methods are usually followed to obtain the final device. In the first one, the bandpass filters are re-optimized within the diplexer environment, which includes the effects of the power divider discontinuities as well as the interaction between the filters. Second methodology the original filter designs are maintained and optimum matching circuit is incorporated into the power divider junction to obtain the required matching in the pass bands. The second approach can be adopted reducing the minimum number of variables for the circuit optimization.

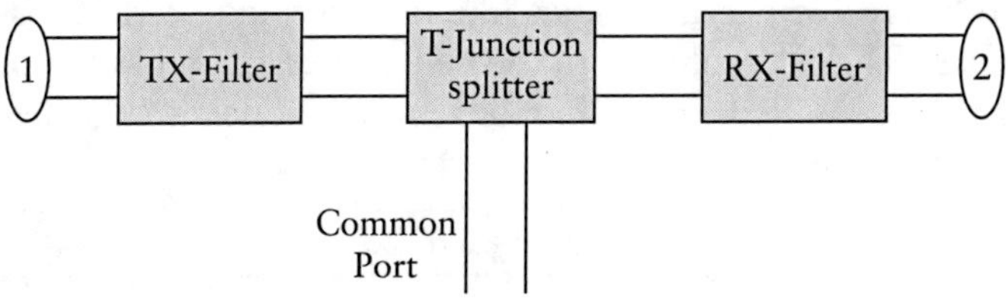

Fig. 6.5: Basic building block of a diplexer employing T-splitter

Design of the diplexer can be divided into three parts: (1) Tx filter, and (2) Rx filter, and (3) T-junction and the overall blocks are shown in Fig. 6.5.

6.2.1.1 Tx Filter Design: Channel filter for the transmitter design working between 12.2 GHz-12.8 GHz is presented in Fig. 6.6. The circuit layout along with electromagnetic simulated response is shown in Fig. 6.7.

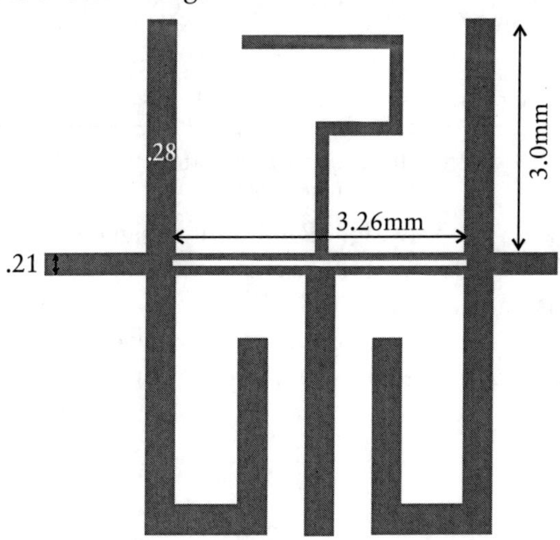

Fig. 6.6: Layout of the Tx filter

The simulated response [9] shows the rejection level of more than 50 dB at the Rx frequency, insertion loss of less than 1 dB. The notch in the main transmission line is introduced (shown as Mod in Fig. 6.7) and analyzed using EM circuit simulator. The notch implemented filter [Fig. 6.6], shows rejection improvement of more than 15 dB at the receive band. There is no change in the other major characteristics like insertion loss, return loss of the filter. This is attributed to the field's interaction in the notch, leading to coupled line behaviour by introducing two nearby poles at the receive band.

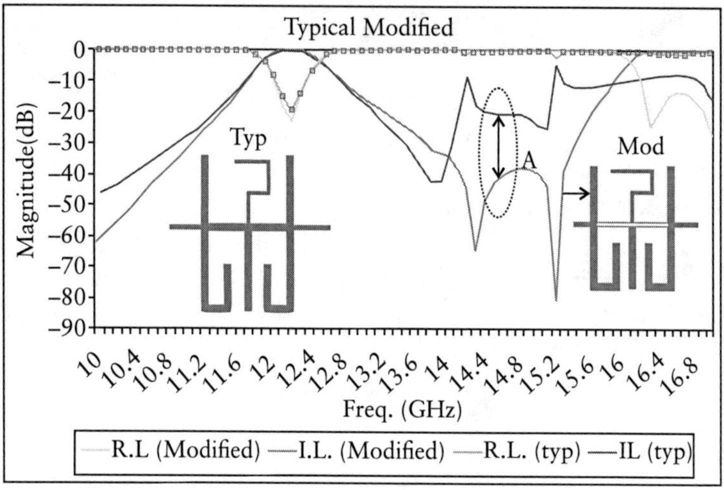

Fig. 6.7: Comparison of the proposed (mod) topology with the standard (Typ) topology

6.2.1.2 Rx Filter Design: Rx filter is designed and optimized at 14.5 GHz. The layout and the current plot of the EM simulated structure is shown in Fig. 6.8. The effect of the notch is also being studied in this configuration. Partial notch in the t-line has no significant affect on the RF performance.

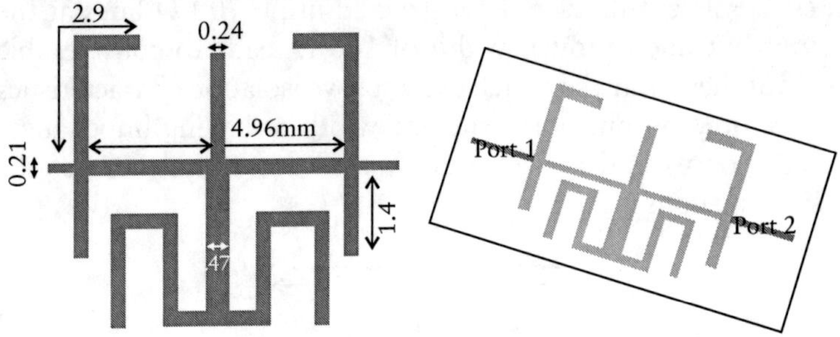

Fig. 6.8: Layout of the Rx filter and current conduction behaviour

Introducing the longer notch [Fig. 6.9], the performance changed considerably. Notch width in the present study is kept constant (0.12 mm) and two different notch lengths are taken in simulation. The study shows the notch introduces poles at the transmit frequency bands which gives around 20 dB improvement. This rejection enhancement overall improves diplexer performance compared to the standard topology suggested by Quendo et al. [11]. As the notch also effects the center frequency, so implementation of notch in receive filter is not implemented as it can affect the layout dimension, which is kept fixed to facilitate testing with the available zig.

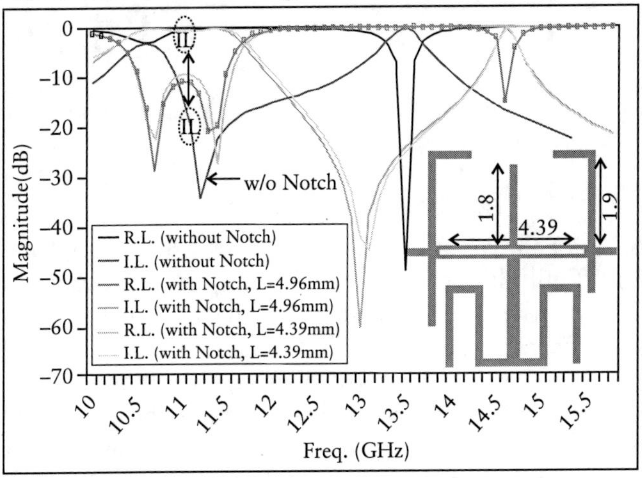

Fig. 6.9: Study of the notch on receive filter performance

6.2.1.3 Power Divider: Power divider is an essential part in the diplexer and T-combiner is chosen to join both the Tx-Rx filter. Lossless T-divider is constructed using 100 Ω lines at the output but due to lower width of 100 Ω leads to considerable losses at the output, so to achieve improve isolation characteristics between the output ports, the line width to having impedances of 70.7 ohms and phases are adjusted accordingly to get the desired response.

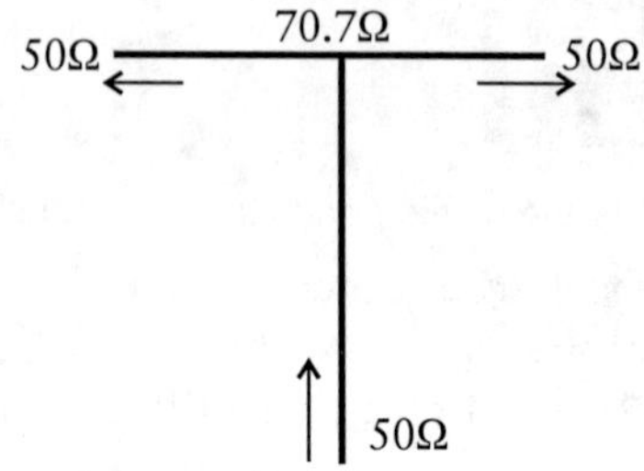

Fig. 6.10: T-combiner topology

Rx-Tx filters are joined together using T-combiner and further optimization has been carried out. The layout of the overall circuit is shown in Fig. 6.11.

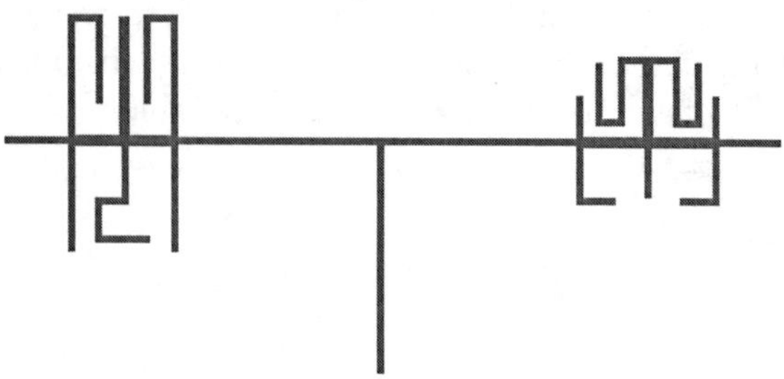

Fig. 6.11: Final layout of the diplexer

6.2.1.4 Fabrication Aspects: The presented filter implemented on the 10 mil alumina substrate having dielectric constant (ε_r) of 9.9 with overall size of 1″×1″. The filter realized on the microstrip substrate and attached using conductive epoxy [10]. Further connectors are attached using the ribbon bonds. To facilitate the measurement, a 2-mm line at the input and the output port is added. The measured result shows pass band insertion loss of less than 3 dB and attenuation of more than 10 dB in the transmit band.

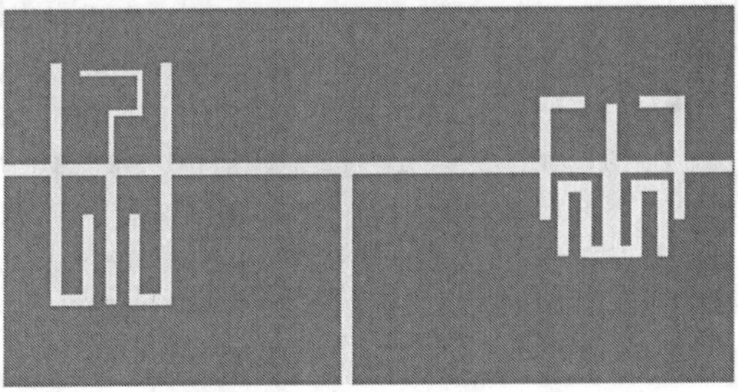

Fig. 6.12: Fabricated diplexer

The results show slight frequency drift with higher insertion loss at the receive frequency band. This is attributed to the fabrication tolerances along with permittivity dispersion. The drift in the permittivity affects the T-combiner matching, which overall degraded the circuit performance. The fabricated diplexer has been measured on the PNA vector network analyzer (8261A) showing the isolation characteristics better than 24 dB. The comparison of the simulated and measured performance is shown in Fig. 6.13.

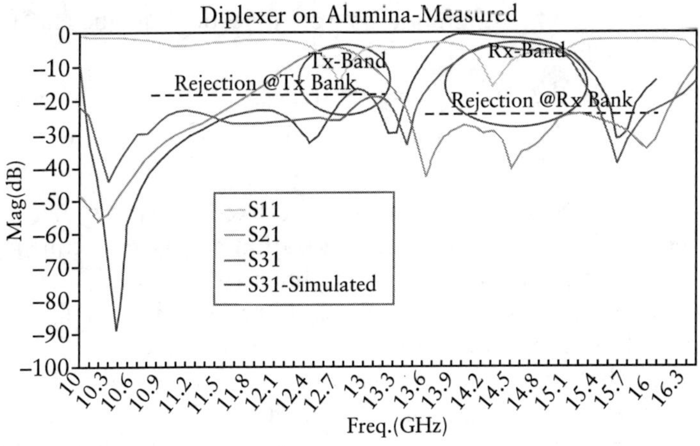

Fig. 6.13: Comparison between the simulated and measured performance

The presented planar approach shows promising result for planar implementation.

6.3 Low Noise Amplifier (LNA)

LNA is the key component in front-end receiver, which dictates its dynamic range. This is achieved by low noise and high gain of the first stage of receiver which is after BPF (band-pass filter) followed by LNA. The main consideration in design of LNA is the trade-off between two important parameters namely noise figure and VSWR [12]. LNA design minimizes the noise figure of its active device by presenting an optimal source reflection coefficient '$\Gamma_{s, opt}$' while output is designed for flat gain and overall stability of the circuit. With an arbitrary source load, 'Γ_s', device yields a noise figure, NF, given by,

$$F = F_{min} + \frac{r_n}{g_s}\left|Y_s - Y_o\right|^2$$

where Ys = g_s+jb_s is the source admittance,

Yo= g_o+jb_o source admittance for minimum noise figure

where r_n = Rn/50 is the equivalent noise resistance usually provided with the noise parameters by device manufacturer. The noise figure of linear two port amplifier can be formulated as a function of only four parameters

(a) F_{min}, the minimum noise figure of the device
(b) Γ_{opt}, the optimum source reflection coefficient for achieving F_{min}.
(c) Rn, the device noise resistance
(d) Zs, the actual source impedance applied to the input of the device

Three of the four parameters are fixed constants that are determined by the device characterization at a given frequency, so only one parameter can be modified to achieve the desired specifications. If $\Gamma_s = \Gamma_{s, opt}$, then above equation changes to F=F_{min}. Designing low noise amplifiers with good input match is main challenge as there is only one load impedance, Z_L that will provide a conjugate output match for a device terminated at the input for minimum noise. However, Zin is generally not equal to Zopt* and mismatch occurs at the input port. So a feedback mechanism or change of Zin is required to make it closer to Zopt* to present minimum noise figure with better matching. In present design, instead of employing standard inductive feedback, a lumped resistor in bias line is introduced. This alleviates problem of grounding of source leads using via or stub which eventually increases circuit size. It also improves input matching without degrading much of the overall gain. The present simulation study is carried out using HEMT device as it has lower noise figure and higher gain due to high transition frequency, f_T. FET has lower noise figure compared to BJT as location of noise generators are different while comparing equivalent circuit.

Single stage of LNA is design is presented using HEMT device. As device is potential unstable in the operating range, so to make it stable some feedback mechanism is required. In the present case, inductance (parasitic) due to leads of device as marked in Fig. 6.14 makes a positive feedback path, thus stability of device increases. The value of inductance is modeled before incorporating in circuit simulator. Further to increase stability of proposed design, resistance in gate bias is introduced [13] which gives dual advantage of high stability with improved matching. Both gate and drain bias circuits are connected closely to HEMT. This is for keeping matching circuit away from bias circuit interference. Bias circuitry uses radial stub due to its broadband characteristics and low insertion loss. Harmonic balance analysis is carried out [9] and it is observed that a single section is not enough to achieve the desired specification of high gain.

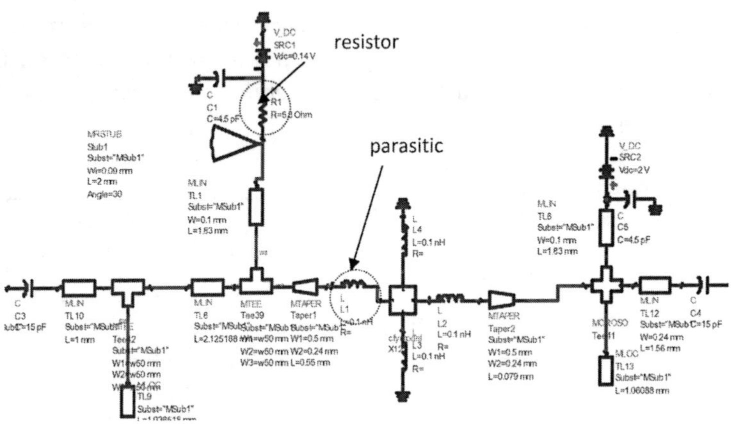

Fig. 6.14: Circuit schematic of single stage LNA

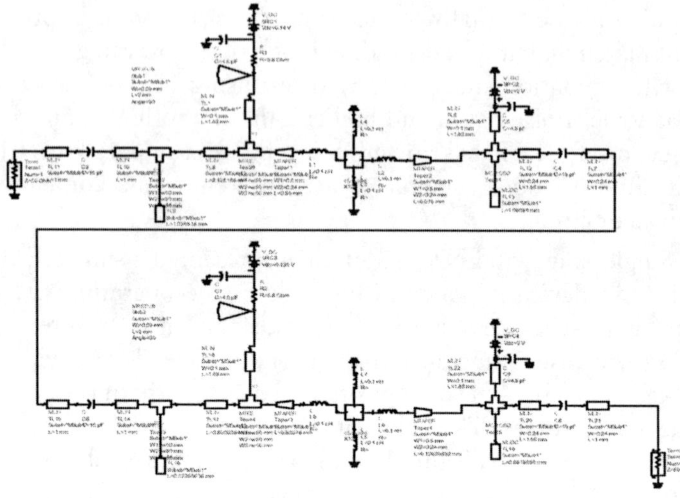

Fig. 6.15: Circuit schematic of cascaded LNA

To enhance gain of transistor, cascaded two stage design is carried out using the standard technique [14]. Figure 6.15 shows the cascaded two stage design of the proposed LNA at Ku-band. Same technique is used in second stage also, and inter-stage matching carried out to achieve higher gain.

Simulated results of above mentioned topology is shown in Fig. 6.16 which shows the optimized performance . Furthermore, there is minimal degradation in gain but resulting stability and return loss is much higher than achievable using standard techniques.

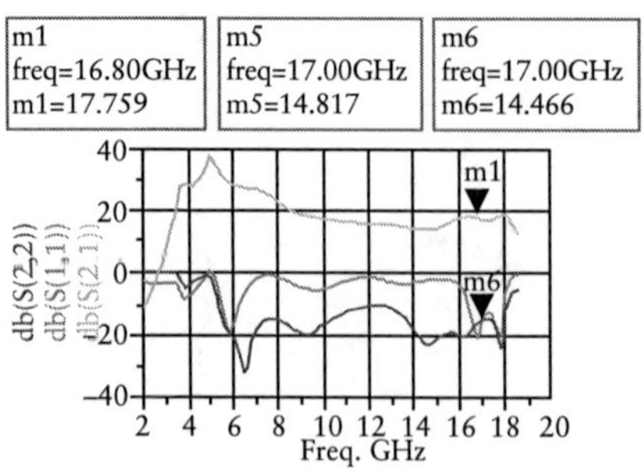

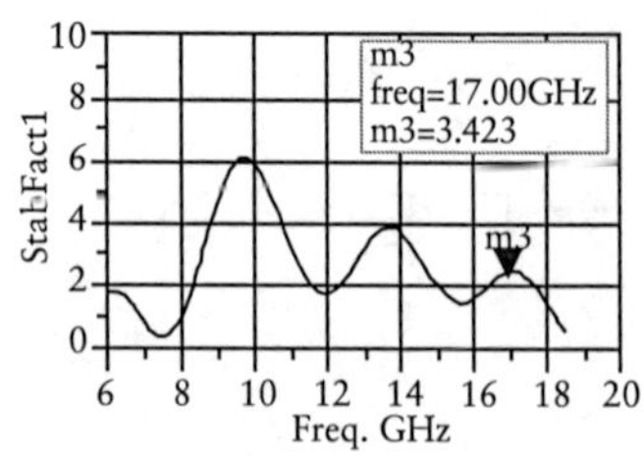

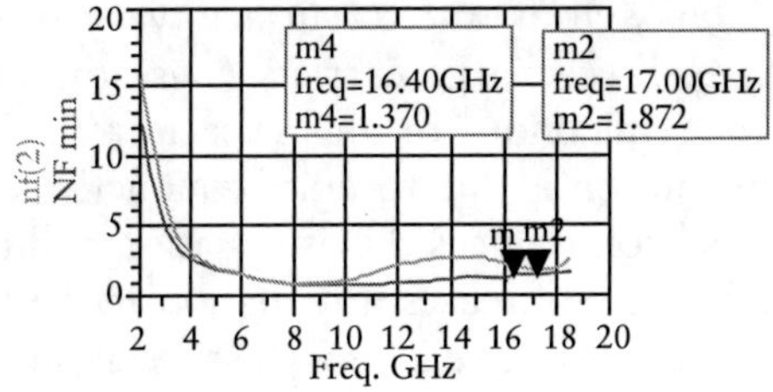

Fig. 6.16: Simulated gain, stability and noise figure of two stages LNA

Another important aspect in the amplifier realization is the consideration of the biasing, which is based on the efficiency, gain, bandwidth, noise, oscillation suppression, power supply so the selection of the quiescent point and holding it constant with temperature and device fabrication to be taken care.

6.3.1 Silicon-based topologies

LNA topologies based on RF-CMOS needs RF characterization of transmission line on silicon substrate [15]. A transmission line is considered as the combination of resistor, inductor and capacitor. Resistance of the line is the combination of the ac and dc component given as (where σ is the skin depth and σ is the conductivity)

$$R = R_{ac} + R_{dc}$$

$$R = R_{ac(gnd)} + R_{ac(trace)} + R_{dc(trace)} + R_{dc(gnd)}$$

$$R_{dc} = \frac{1}{\sigma_0 \sigma_r \omega t}$$

$$R_{dc} = \frac{1}{\sigma_0 \sigma_r \delta} \qquad \text{... (6.1)}$$

$$C = \frac{(\varepsilon_r l)}{60c \ln\left(\frac{8h}{w} + \frac{w}{4h}\right)}; \ \frac{w}{h} < 1$$

To implement the RF circuits on standard CMOS substrate various aspects to be considered as silicon substrate is considered neither a good conductor nor a good dielectric due to finite conductivity and resistivity. This is due to the fact that the contribution of the metallic losses is not affected by the substrate resistivity and major loss contributions are from dielectric as well as due to surface wave propagation in silicon substrate. Free charge carriers cause the main part of the losses in the semiconductor and as seen from Eq. (6.2), the main loss contribution is due to finite conductivity. Employing high resistivity silicon reduces the attenuation and one order of increase in resistivity decreases attenuation by half in C-band. The cut off functional frequency for passive circuits is given as:

$$f_c = \frac{1}{2\pi\rho\varepsilon} \qquad \text{... (6.2)}$$

where $\rho = 1/qn\mu$ (q: charge concentration, μ: mobility)

As evident from above, transition frequency depends on the resistivity ($\rho=1/\sigma$) which is determined by the doping level. Role of conductivity is an important parameter in semiconductor which is dependent on concentration and mobility in contrast to conductor. In case of silicon substrate, the losses are primarily from substrate and interface losses induced between the insulation layer and silicon substrate [7]. These losses can be minimized by employing high quality gate oxide in conjunction with high resistivity silicon substrate. Substrate losses for silicon in case of 10k Ω-c-m are 0.032 dB/cm which increases to 32 dB/cm for the resistivity of 10 Ω-c-m. High resistivity silicon substrate (1 k Ω-cm) is having attenuation of the order of 10^{-3} compared to 10^{-4} for silicon with resistivity of 10 k Ω-cm. Metallization thickness of 10 um makes skin depth comparable with the thickness and cause signal degradation. Conductor attenuation can be written as:

$$\sigma_c = 0.072\frac{\sqrt{f(GHz)}}{wZ_0}\lambda_g dB/\lambda_g = \frac{8.686}{Z_0 w}\sqrt{\frac{\pi f \mu}{\sigma}} dB/m \qquad \text{... (6.3)}$$

In terms of quality factor it can be represented as.

$$\alpha_c = \frac{8.68\beta n}{Q_u}(dB/mm);\ where\ Q_u = \frac{Q_L}{1-10^{-\left(\frac{IL}{20}\right)}} \qquad \text{... (6.4)}$$

where $\beta=2\pi/\lambda_g$, n-number of half-wavelengths

Subsequently, quality factor associated with attenuation constant can be written as:

$$Q_c = \frac{27.3}{\alpha_c} \approx 0.63h\sqrt{\sigma f(GHz)}$$

Dielectric attenuation constant attached with the loss can be expressed as

$$\alpha_s = 27.3\frac{\varepsilon_r(\varepsilon_{eff}-1)}{\varepsilon_{eff}(\varepsilon_r-1)}\frac{\tan\delta}{\lambda_g}dB/\lambda_g \text{ where } \lambda_g = \lambda_0/\sqrt{\varepsilon_{eff}} \quad \text{... (6.5)}$$

So total attenuation can be written as $\alpha = \alpha_c + \alpha_s$. Further for silicon $\alpha = \alpha_c + \alpha_L + \alpha_d$

Subsequently, quality factor associated with attenuation constant can be written as [1]:

$$Q_s = \frac{27.3}{\alpha_s} \approx \frac{1}{\tan\delta_d + \tan\delta_L}; \quad \frac{1}{Q} = \frac{1}{Q_c} + \frac{1}{Q_s}; \quad Q = \frac{\beta}{2\alpha}$$

The losses in the silicon substrate are slightly higher due to finite resistivity which results in higher noise. The reason is the lower quality factor associated with spiral inductors where inductor loss resistance contribute in the overall noise figure.

RF modelling for standard MOS models are not allowed as terminal resistors, bulk effects and non-quasi static plays important role at higher frequencies.

6.3.2 CMOS-based LNA

RF-CMOS circuits are highly linear and can be integrated with the digital circuit resulting in the implementation of the single chip. Various input matching topologies are (a) resistive terminations (b) shortend gate (c) shunt feedback (d) inductive degeneration. Inductively degenerated input stage provides the lowest noise figure under matching conditions. The main contribution of the noise in CMOS device is the noise generated by the distributed resistance of the poly-silicon gate. Inductive degeneration topology is preferred choice as it create resistive input impedance and also provides lowest noise figure. Bond wire for the ground connections introduces inductance and resistance which directly changes the source impedance resulting in lowering of effective transconductance and efficiency which can be minimized by having multiple pads and bond wires. The concept of employing two stages (NMOS and PMOS) allows the re-use of the bias current resulting in lower current consumption [13]. The transistor is biased close to f_t for lowering the NF as at higher gate bias voltage the f_t drops due to higher mobility degradation. The input and output coupling of the LNA to be reduced so as to prevent oscillations.

The noise figure for the resistive terminations can be simplified as

$$NF = F_{\min} + \frac{NG_n}{R_s}\left|R_s - \frac{Z_{opt}}{N}\right|^2 = 2 + \frac{P_n}{(kTR\Delta f)} \quad \text{... (6.6)}$$

where P_n is the noise power at the input, R_s is driving source resistance, G_n is noise conductance and Z_{opt} is the optimum noise impedance. Gate resistance noise is proportional to the width and inversely proportional to the length of the device. Selection of the optimum device width, bias voltage at the selected frequency results in better input match and minimum noise figure. The 0.18 um technology is having f_t of around 60 GHz and as the gate length scales down the subsequent noise also reduces. The main advantage of CMOS-based approach is the single chip solution [16].

CMOS devices exhibit higher linearity than other topologies so it becomes ideal choice below 10 GHz. In case of CMOS-LNA, the device transconductance is constant with V_{gs} which show increased level of linearity.

Present trend having GaN high-electron mobility transistors (HEMTs) for realization of robust low noise amplifiers due to high breakdown voltage and low noise characteristics. These results in elimination of the limiter circuits which protect the entire system and realization in smaller volume is possible.

6.4 RF Amplifier Design

The amplifier design is based on the various characteristics such as frequency, bandwidth, power gain, gain flatness, return loss, stability, efficiency, inter-modulation performance, dynamic range. The RF amplifier with the centre frequency at Ku-band designed is shown using the DC block topology in above incorporation of HEMT transistor has been used for the realization of an amplifier is shown in this section. The stability and the input/output reflection coefficient has been found out using equations 6.7-6.10 [5, 6]. The device shown stability in the broad frequency range (Rollett Stability factor K>1 and lDl<1). The conjugate matching concept can be applied for the maximum gain given by equation 6.9[12].

$$\Gamma_{in} = S_{11} + \frac{S_{12}S_{21}\Gamma_L}{1 - S_{22}\Gamma_L} \quad \Gamma_{out} = S_{22} + \frac{S_{12}S_{21}\Gamma_S}{1 - S_{22}\Gamma_S} \qquad \text{... (6.7)}$$

$$K = \frac{1 - |S_{11}|^2 - |S_{22}|^2 + |D|^2}{2|S_{12}S_{21}|} > \quad D = |S_{11}S_{22} - S_{12}S_{21}|^2 < 1 \qquad \text{... (6.8)}$$

$$G_A = \frac{1 - |\Gamma_S|^2}{|1 - S_{12}\Gamma_S|^2} |S_{21}|^2 \frac{1 - |\Gamma_L|^2}{|1 - S_{22}\Gamma_L|^2} \qquad \text{... (6.9)}$$

For high gain reflection coefficient are selected as

$$\Gamma_S = (\Gamma_{in})^* \qquad \Gamma_L = (\Gamma_{out})^* \qquad \text{... (6.10)}$$

$$\Gamma_S = S_{11} + \frac{S_{12}S_{21}\Gamma_L}{1 - S_{22}\Gamma_L} \qquad \Gamma_L = S_{22} + \frac{S_{12}S_{21}\Gamma_S}{1 - S_{11}\Gamma_S}$$

where Γ_s-load reflection coefficient, Γ_s-source reflection coefficient

The bias lines are incorporated with the high and low impedance section. Bias is adjusted for the Class A operation with Ids=15 mA with Vds=2.0 V. The non-linear device model of HEMT transistor (Materka model) is incorporated. So as to perform linear and non-linear analysis of integrated amplifier. Circuit is realized on 10 mil alumina substrate employing microstrip dc block at the input and the output of the amplifier. The circuit is attached to the Kovar plate using epoxy and connectors are soldered. The accuracy of the line width and spacing of the DC block is maintained with high accuracy while etching using wet chemistry. A wide band DC block with better than 1 dB insertion loss and 12 dB return loss from 14-20 GHz is realized with coupled line have precise gap separation.

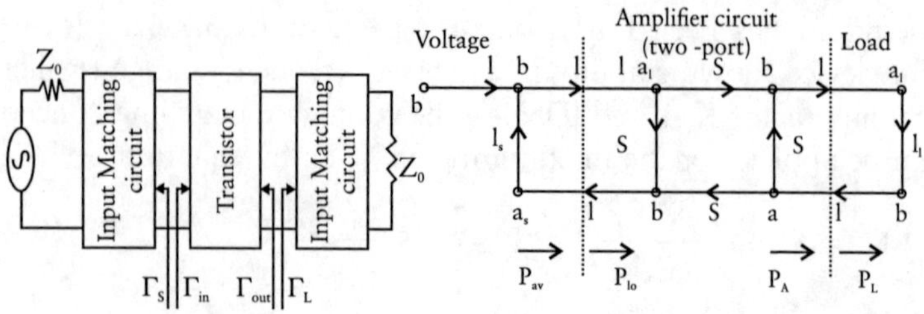

Fig. 6.17: Schematic for amplifier matching

6.4.1 Experimental Results

Fabricated circuit is characterized where measured result shows 1-dB compression at 2 dBm input power with the gain of 9 dB. Fig. 6.18 shows simulated result of non-linear analysis and resulting output power of different harmonics and Fig. 6.19 shows simulated gain and return loss from standard software. The full layout is shown in Fig. 6.20.

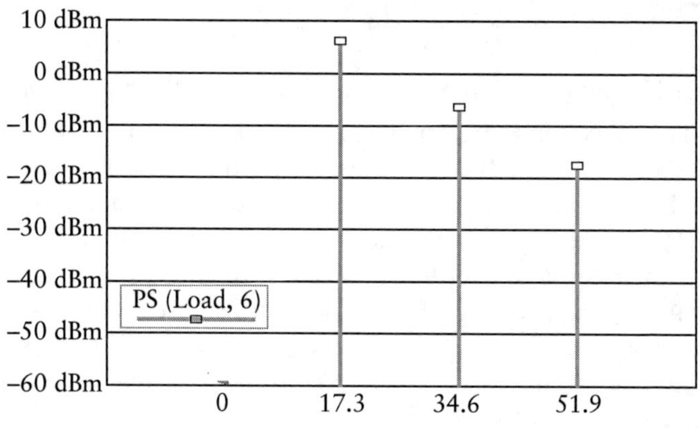

Fig. 6.18: Simulated output power

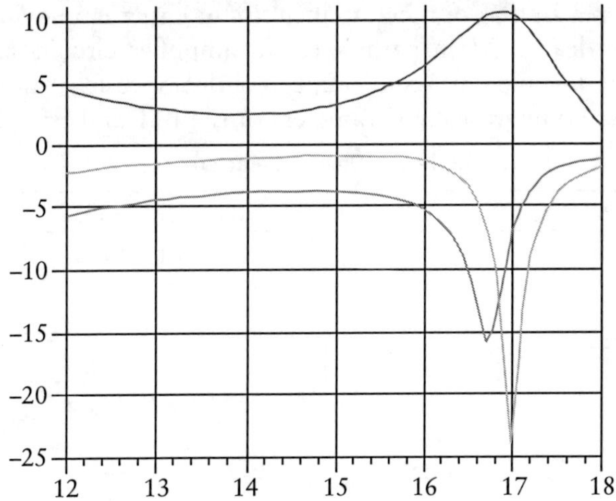

Fig. 6.19: Gain plot

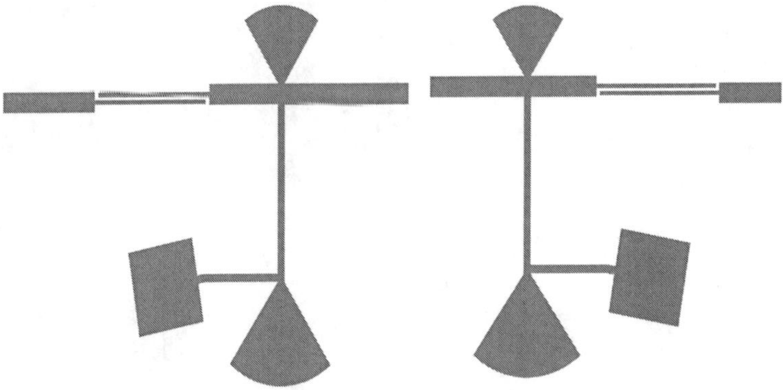

Fig. 6.20: Schematic of the RF amplifier

Designing RF sub-system, the first amplifier is a low noise amplifier having considerable gain for the noise control and the last amplifier has the highest third-order intercept point so as to set the dynamic range. SAW bandpass filter is employed to eliminate unwanted frequency components whereas incorporation of attenuator improves matching.

Present trend of implementing amplifier is based on the CMOS technology to have SOC concept and selection of BJT or FET device become important consideration. Apart from CMOS, BiCMOS and SiGe-BiCMOS techniques are being employed for the realization of the amplifier circuit either employing BJT or FET device. Main parameters in amplifier circuits are the transconductance (g_m) and output conductance (g_o). Table 6.1 provides comparatively parameters of the BJT and FET device.

Table 6.1: Device characteristics

Parameters	BJT	FET
Transconductance (gm)	$\frac{qI_c}{kT}$	Constant $\times\sqrt{I_d}$
Cut-off frequency (f_T)	$f_T = \frac{g_m}{2\pi(R_E C_{BE} + R_E C_{BC} + \tau_B + \tau_C)}$	$\frac{g_m}{2\pi(C_{BS} + C_{GD})}$
Maximum frequency of operation (f_{max})	$\sqrt{\frac{f_T}{8\pi C_{CB} C_B}}$	$\frac{f_T}{2\sqrt{(R_I + R_g + R_s)\, g_{ds} + 2\pi f_T R_G C_{GD}}}$
1/f noise	Depends upon emitter-base junction	Si/SiO interface
General model	Gummel poon	Curtice, TOM, Materka

Transconductance demonstrates ability of device to source the current whereas output conductance shows sourcing of the current and swings in the output voltage. For the same biasing condition, BJT achieves three times transconductance and three times drive capability compared to FET, so incorporated at lower end of the microwave band.

Compared to BiCMOS process, SiGe BiCMOS process leads to higher speed, better base and collector transit time, improvement in f_T and lower parasitic. Lower parasitic and higher g_m (transduction) leads to better noise figure.

6.4.2 Power Amplifier

Standard power amplifier employs FET-based devices at higher frequencies to achieve the desired output power. The approach of single stage to achieve desired output power yields thermal as well as reliability issues.

Alternatively, chain concept, corporate or radial approach is implemented to achieve desired output power by combining two or more transistors yielding better input and output impedance match with ripple free gain characteristics. Power amplifier can be operated in low, medium and high power region whereas linear and medium power ranges is having linear operation (class A) and high power region is operated at 1-dB compression point (class B). The efficiency is dictated by the class of operation and specific to the application (linearity, inter-modulation, interferences). Class D, E, J are supposed to achieve higher efficiency but they are operated in switched mode and needs resonators at the output. So class AB operation is carried out to obtain high efficiency and better linearity. Bias circuit also plays an important role as excessive gate current shorten the device life and to avoid such phenomena appropriate gate resistance to be considered for achieving linearity and power performance. RC networks at the bias are to be employed to prevent low-frequency oscillations. A combination of diode compensation in the bias network and thermopads are used to control the gain variation over the temperature. Driver stage preceding the power amplifier is designed to provide output drive having substantial gain. This is done by having conjugate matching at the input and output of the device based on the small signal design approach. Dies are cascaded in the driver amplifier stages to achieve the desired performances. In power amplifier, the device is drive into compression (1-dB compression) so as to achieve power gain and output power at the desired frequency to achieve higher PAE [18-19].

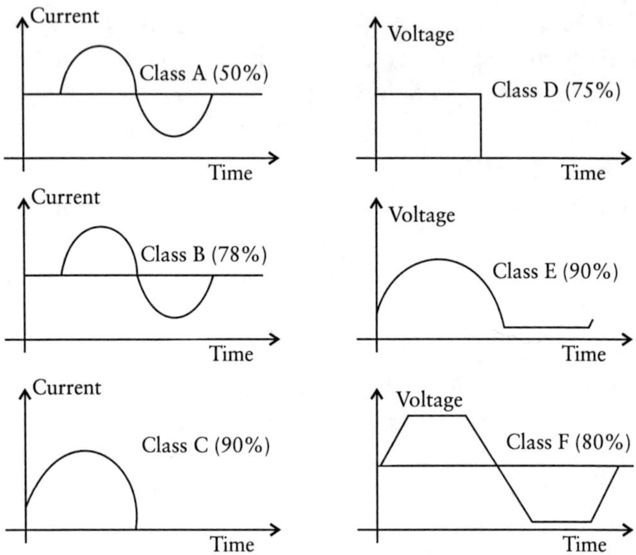

Fig. 6.21: Power amplifier waveforms and efficiency

Large signal S-parameters are used to obtain the desired output power and in case of unavailability of the same load and source pull measurement techniques are employed to achieve the same. Further the choice of substrate also plays an important role so as to carry the power from the device. The maximum average power which can be handled is given as

$$P_{m} = \frac{\theta}{\Delta T} \quad \text{... (6.11)}$$

where ΔT is the temperature rise above ambient and is inversely proportional to K_{TH}.

The rise in channel temperature can be represented as

$$T_j = R_{\text{th}} P_{\text{diss}} + T_c \quad \text{... (6.12)}$$

where $$P_{diss} = DC_{(i/p\ power)} + RF_{in} - RF_{out}$$

where R_{th} is junction thermal resistance, P_{diss} is average dissipated power, T_c is case temperature [1, 19].

Considering device thickness to be small compared to dimension of the device, the heat flow in the vertical column can be represented as R_{th} for the device thickness of 'T' can be written as

$$R_{th} = \frac{T}{\sigma_{th} \times \text{Area}} \quad \text{... (6.13)}$$

where R_{th} is vertical column thermal resistance, σ_{th} is thermal conductivity of the material.

GaAs topologies implemented in the power amplifier is having low efficiency resulting in thermal issues, increased size and cost due to auxiliary circuits. The device operated in the saturation region shows high power dissipation and to be carefully placed in the package so as to avoid heating of nearby components. The current efficiency of the power amplifier is limited and further improvement is not possible due to device limitation. Thermal imager plot shows (Fig. 6.22) the generation of higher level of heat in the device and surroundings compared to other portion.

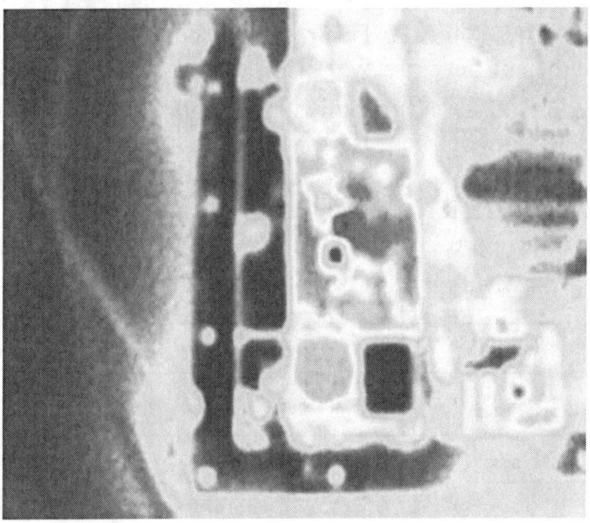

Fig. 6.22: Thermal imager of the power amplifier device

This necessitates the incorporation of the higher bandgap material such as GaN device in the system.

6.4.3 Selection Criteria and Construction

Selection of the device is most important criteria to achieve the optimum and desired performance. GaN HEMT device is having large material band gap which permits them to be operated at high drain voltages and high power density within small size. Basically GaN device on Si and SiC are employed keeping into considerations various parameters as tabulated in Table 6.2.

Table 6.2: Major specifications for various bands

Parameters	Si	SiC
Heat Conduction	Good	Very good
Large aperture conversion	Fair	Very good
Crystal quality	Very good	Fair
Growth layer quality	Very good	Fair
Current density	Low	High
Cost	Low	High

Heat dissipation in SiC is faster compared to Si as thermal expansion coefficient of SiC is 4.6×10^{-6}/°K matches with GaN which is having an average expansion coefficient of 4.9×10^{-6}/°K from 25° C to 1000° C. This results in better cooling of device due to efficient heat removal [20].

The key consideration for the development of GaN HEMT is the selection of the base material. A crystal wafer such as sapphire, SiC or Si is employed as a base substrate and GaN along with AlGaN are formed sequentially using MBE or MOCVD as shown in Fig. 6.23.

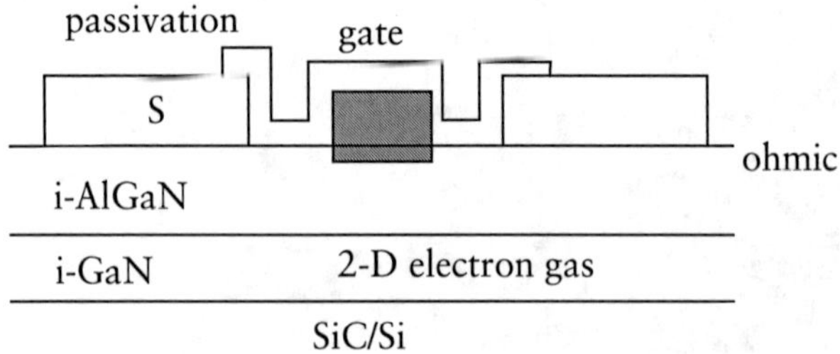

Fig. 6.23: Generic structure of GaN device

GaN HEMT also having excellent FOM (figure-of-merit) shown as Rds × Qg making it better for high frequency of operations. The dislocation density of the HEMT structure as shown above on Si and SiC are $8\times10^{8}cm^{-2}$ and $2\times10^{8}cm^{-2}$ respectively and after prolonged annealing the samples on SiC are unaffected which indicates their superiority in high temperature and high-power applications [15].

Table 6.3: Wide band material properties and associated advantages

Material Property	Device Operation	FOM	Advantage
High breakdown filed	High voltage	Power density	Smaller die
	High doping	Power gain, IMD, output impedance, efficiency	Lower energy usage
High thermal conductivity Wide bandgap	High temperature	Smaller die size More power/ die	Smaller package
High electron velocity	High frequency	High f_T, high f_{max}	High system frequency

Realization Approach

The main steps for the realization of the power amplifiers are:

- Selection of the device
- Selection of bias and biasing arrangement
- Optimum source and load impedances
- Power contours at desired frequencies
- Transformation for raising load resistance to 50 Ω
- Input and output matching network.

GaN HEMT devices are operated at higher voltages which increase device optimum impedances and lowers the device parasitic capacitances for a given output power capability. Alumina is chosen for realization of the compact structure with better RF performance. Sample architecture (Fig. 6.24) shows implementation of power amplifier module lower input and output doctor/combiner with cascading of devices.

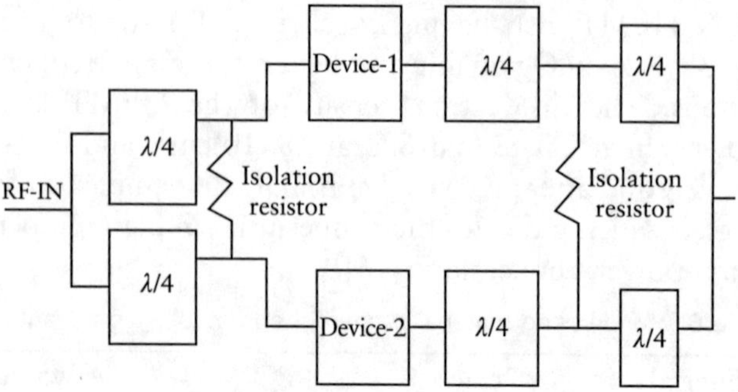

Fig. 6.24: Simple architecture

Conventional branched line hybrid [1] suffers from limited bandwidth and, in addition, the length variations due to fabrication tolerances lead to inaccuracies at higher frequencies. Also quarter-wavelength transmission line in series/shunt arms results in performance deterioration at higher frequencies due to parasitic reactances associated with junctions and unwanted interactions between strip conductors. This can be circumvented by utilizing circular or rectangular disk-shaped four-port planar circuits. Compared to circular disk topology, rectangular disk is chosen for low loss and less complexity. Inherent limitation of planar topology is its narrow band operation [22] and to circumvent the losses and consequent performance degradation, associated with these topologies, unequal open circuited stub symmetrically at the series arm is introduced with diagonal feeding at optimum angle. This approach allows changing the overall impedance matrix resulting in flat coupling response at higher frequency bands.

Fig. 6.25: Disk-based multiport topology

Planar power divider and combiners are the basic building block in microwave circuits employed for communication purpose. The branch line and Wilkinson power divider are the generic circuits for this purpose where 90° and 0° phases are generally obtained [1]. The multiport operations from the same topology can be obtained by cascading the units which results in the enhanced size whereas incorporation of the lumped resistors in Wilkinson topology increases the circuit complexity. Multiple ports by cascading the same units yields even number of ports whereas to achieve odd no ports (typically 3) are also proposed using structure as proposed by Chiu et al. [23]. In contrast, multiport divider and combiner topology using circular sector-shaped planar components are proposed for antenna applications [4] which are having multiple output ports in even and odd configurations without cascading. The sector-shaped geometry is having compact size, equi-phase and non-attachment of external components. The added advantage of these topologies is the flexibility in obtaining unequal output power at various ports which can be controlled by adjusting the locations and widths of the microstrip lines [24-25].

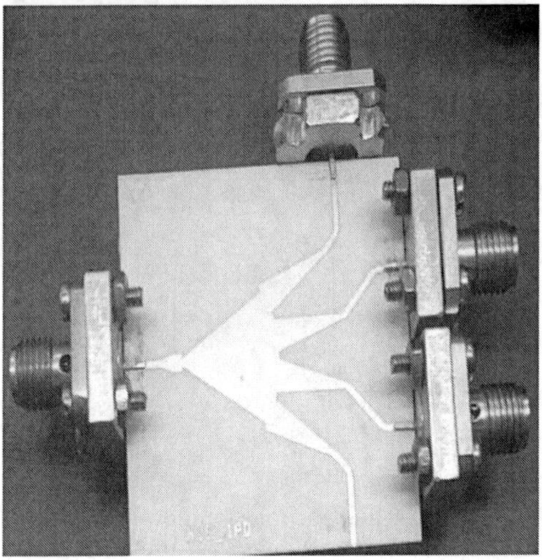

Fig. 6.26: Disk-shaped topology for cascading multiple devices

Patch coupler can be implemented for the design of the output circuits of the PA. Disk coupler can realize impedance conversion and also harmonic impedance control realization by the selection of the impedance conditions of terminal ports.

Disk-based topology (α~90°) which is planar, simpler having considerable size reduction along with cost effectiveness. Also the assembly is fully scalable and can be tweaked for the required frequency ranges and can be employed for the combiner and array antenna applications [24]. Recently, compact single and dual-band patch filtering 180° hybrid couplers based on circular patch resonator is presented by utilizing the electric field distributions of the degenerate resonant modes in the unslotted/slotted patch resonators [41].

Power divider and combiner with isolation resistor as well as mismatch at lower end of frequency to be carried out to curtail oscillations and achieve optimum performance as shown in Fig. 6.27 with the selected internally matched device (IM GaN). Single device is characterized with different input levels and loads to find out the optimum conditions without using the load-pull technique followed by the arrangement as shown in Fig. 6.27. The overall concept drastically reduces the need for multiple stages while meeting the desired specifications [26].

In comparison to traditional topology main advantage associated with GaN approach is achieving higher power density with a single chip whereas equivalent-power GaAs amplifier comprises multiple chips to achieve the same functionality (Fig. 6.27).

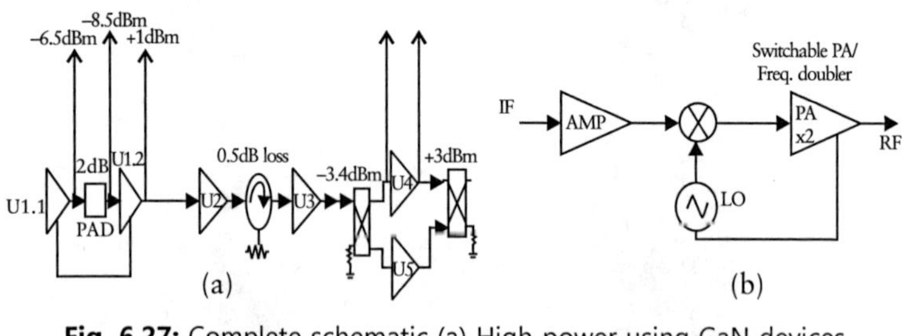

Fig. 6.27: Complete schematic (a) High power using GaN devices (b) Switchable power amplifier using doubler

The body network in which the body of the CS amplifier at the power stage is connected to the drain through a large resistance is very effective to linearize the PA. The linearity improvements are characterized with intermodulation distortions (IMD3) and AM–PM measurement.

Si-Ge based transistor

SiGe devices can be attractive option in RF and microwave domain. The advantage of silicon wafer processing is the higher range of integration and well-controlled process. Also due to bigger wafer size the cost effectiveness is also an added advantage.

The small amount of Ge in Si leads to hetrero-junction and results in higher efficiency. Varying Ge concentration modify electric field across the base improving the carrier concentration which improves flatness of the output characteristics of the device. Also breakdown voltages are low for high speed digital applications beside relatively cheaper. SiGe devices for low power applications are well suited for realization of complete chip on Silicon.

6.5 Dielectric Resonator-based Oscillator (DR Oscillator)

Oscillator is one of the key components in communication systems. Present-day technology of using low frequency oscillator followed by multiplier stages have become obsolete due to requirement of large real estate and lot of tuning required. So, developing a high frequency oscillator makes system more reliable and compact. This necessitates a proper modelling and extensive non-linear simulation. Oscillator can be LC-type (SAW, DR, Crystal based) with much higher Q or RC-type (ring, relay type) with lower Q. DR (dielectric resonator) is high Q ceramic resonators conveniently sized for MIC applications. DR offer integration capability with MICs as well as quality factor and frequency stability approaching cavity resonators appropriate for high Q requirements. DR sustains same electromagnetic field modes as of metal cavity resonators. DR application in active and passive circuits has been rapidly increasing due to small size, low price, insensitivity to mechanical vibrations, insensitivity to electrical power transients, tone free output, direct operation without multiplier stages and excellent integrability in MICs [27]. Dielectric material generally barium tertatinanate having cylinder shape provides the properties similar to the cavity resonator.

The fields in the DR puck extends beyond the material which is responsible for the tuning. The change in the position of the DR puck results in the increase or decrease in the inductance. Detailed non-linear analysis at C-band and methodology of modelling and implementation is presented in this section.

There are commonly two topologies of DR-based oscillator: reflection type oscillator and feedback oscillator. Feedback oscillator topology is preferred due to its insensitiveness from load pulling and pushing. There is commonly two ways of implementing the DR puck in the circuit: as a passive stabilization element Stabilized DRO or as a circuit element in a frequency determining networks (Stable DRO). A stable DRO consists of an active device and a passive frequency determining element (DR puck). The material for the resonator is chosen according to the frequency requirements [28].

Diameter (D = 2a) of a cylindrical DR is chosen for $TE_{10\delta}$ mode such that

$$\frac{5.4}{\left(\beta \times \sqrt{\varepsilon_s}\right)} > 2a > \frac{5.4}{\left(\beta \times \sqrt{\varepsilon_r}\right)}$$

where ε_s = relative permittivity of the substrate,

ε_r = relative permittivity of a DR,

β = free space propagation constant

The radius can be determined 'a' using the formula

$$a = 1.2025 \times \frac{c}{\omega_0}\left[\frac{1}{\sqrt{\varepsilon_r}} + \frac{1}{\sqrt{\varepsilon_s}}\right]$$

The height of the puck should be numerically equal to D/h ≈ 2.53 and the upper cover height should be 0.8h (to avoid spurious modes), where h is the height of the substrate.

DR puck is chosen considering following characteristics:

- Q-factor, which is approximately equal to the inverse of the loss tangent,
- Temperature coefficient of the resonator, and
- Dielectric constant.

6.5.1 Modes in Dielectric Resonator (DR)

Both H and E resonant modes can be excited in a dielectric resonator [29]. The commonly used mode in cylindrical resonator is $TE_{10\delta}$. The $TE_{10\delta}$ mode of cylindrical DR has the highest Q among all resonant modes. The $TE_{10\delta}$ mode also operates like a magnetic dipole mode with an efficient magnetic coupling to a microstrip line (Fig. 6.28).

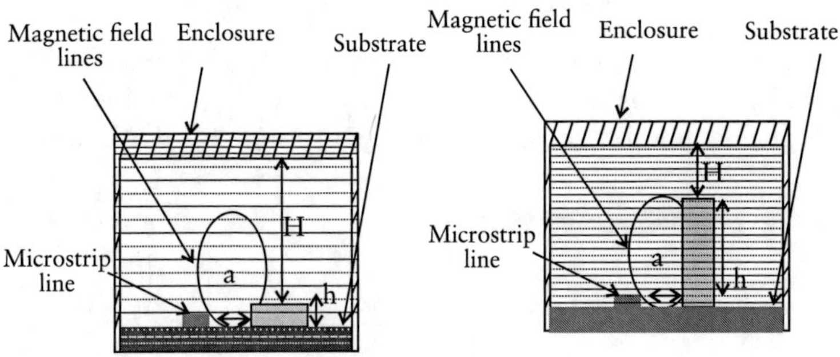

(a) Lateral implementation (b) Transverse implementation

Fig. 6.28: Coupling mechanism of dielectric puck with microstrip line

6.5.2 Modelling of DR

A DR coupled to a microstrip line under specified shielding condition can be modelled as parallel RLC equivalent circuit. The equivalent circuit and its implementation in CAD tool is shown in Fig. 6.29. The coupling coefficient β is expressed in terms of S_{110} and S_{210}, which represent the reflection and transmission coefficients respectively. The S matrix at an angular resonant frequency of the $TE_{01\delta}$ mode is expressed as

$$[S(\omega_0)] = \begin{bmatrix} \frac{\beta}{\beta+1} & \frac{1}{\beta+1} \\ \frac{1}{\sqrt{\varepsilon_r}} & \frac{\beta}{\sqrt{\varepsilon_s}} \end{bmatrix}$$

As insertion loss $L_0 = -20\log(S_{210})$, so coupling coefficient = 20 log (S_{210}), so coupling coefficient β can be determined from

$$\beta = \frac{R}{2Z_0} = \frac{S_{110}}{S_{210}} = \frac{S_{110}}{(1-S_{110})} = \frac{(1-S_{210})}{S_{210}} = 10^{\frac{L_0}{20}} - 1$$

$$\Gamma = \frac{\beta}{\beta+1};\ R = 2\beta Z_0;\ \ L = \frac{R}{\omega_0 Q_u};\ C = \frac{1}{\omega_0^2 L}$$

Q_u = unloaded quality factor $= \dfrac{\omega - \omega_0}{\omega_c} = \dfrac{\beta}{Q_E} = (1+\beta)Q$ where ω_0 = angular resonant frequency

$$Z_{in} = 1 + \frac{2\beta}{1 + jQ_u}$$

A simple transmission line is fabricated and DR puck having permittivity 30 is chosen for the analysis. Values obtained from the given equation can be directly incorporated in CAD tool.

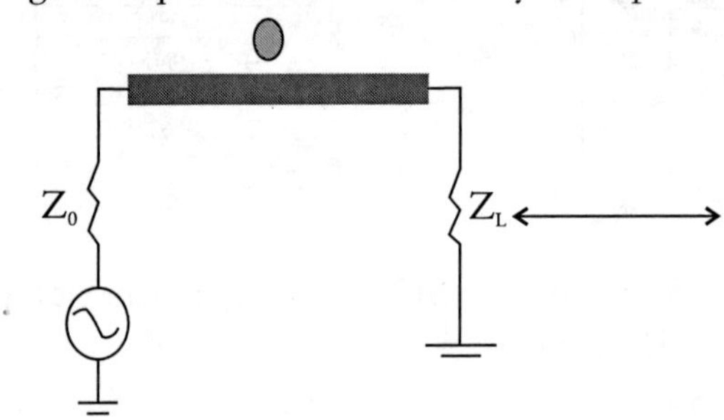

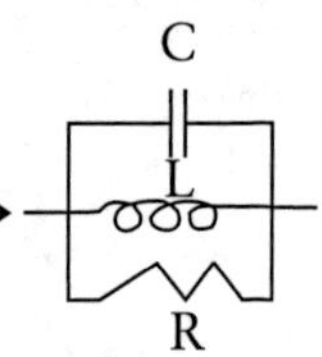

Fig. 6.29: TL interaction with DR **Fig. 6.30:** Equivalent circus

Measured result shows the return loss of 20 dB, insertion loss of 2 dB, and bandwidth of 150 MHz. From the measured results, calculated parameters are:

β (coupling coefficient) $\approx S_{11,\ C}/S_{21,\ C}$ = 6.45, Γ (reflection coefficient) = $\beta/(\beta+1)$ =.86

$R=2\ \beta\ Z_0 \approx 645$ $C= 1/(\omega^2 L) \approx 365$ pF $L=R/(\omega_0 Q_U) \approx 1.36$pH

$Q_U = fc/B.W \approx 154.3$ where $Q_U = Q_L(1+\beta)$ $\Gamma = N^2R/(2Z_0+N^2R)$ so N = .95

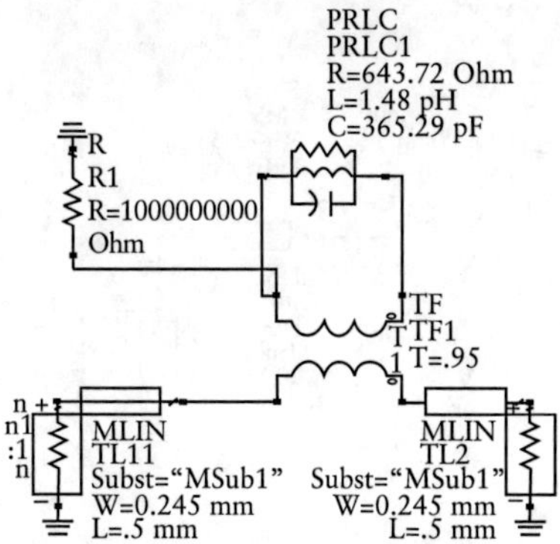

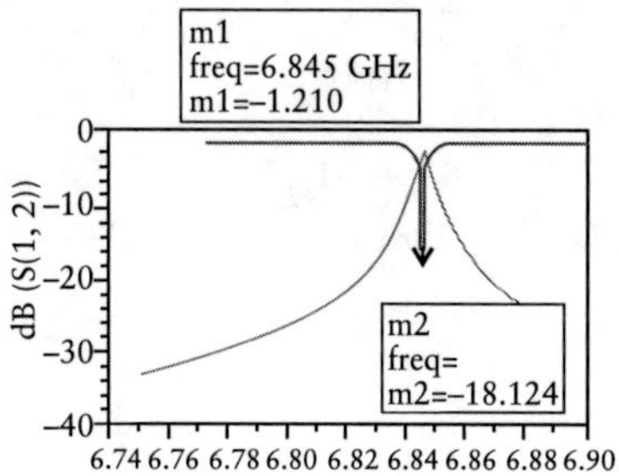

Fig. 6.31: Complete schematic **Fig. 6.32:** Simulated plot of the equivalent circuit

The non-linear simulation has been carried out using harmonic balance simulator [9]. The HEMT transistor (for better phase noise) has been chosen in a common source topology. First cut oscillation frequency is determined by small signal analysis. The DR model (Fig. 6.30-Fig. 6.32) has been incorporated as the frequency determining resonator circuit. The full circuit (Fig. 6.33) is subjected to harmonic balance analysis. The Vds is kept 2.6 V, corresponding to Ids of 15 mA. The convergence of the circuit has been attained using appropriate setting of the Samanskii constant and keeping simulation in Kyrlov mode. Results are plotted as shown in Fig. 6.34. The integrated DRO shows the oscillation frequency of around 6.25 GHz. The shift of 600 MHz in the DR resonance and the oscillator frequency is due to radiation loss which is not taken into account, affecting the Q of the circuit. This way the effect of circuit parameters, model parasitics can be easily seen giving more confidence to the designer.

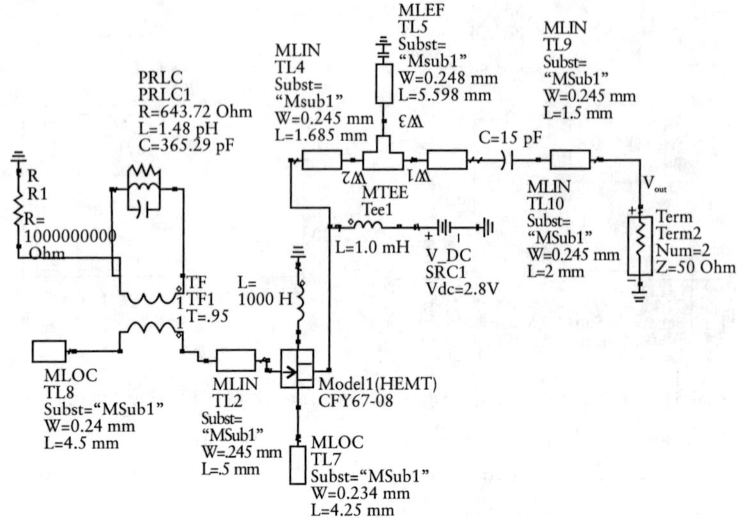

Fig. 6.33: Schematic of the DRO

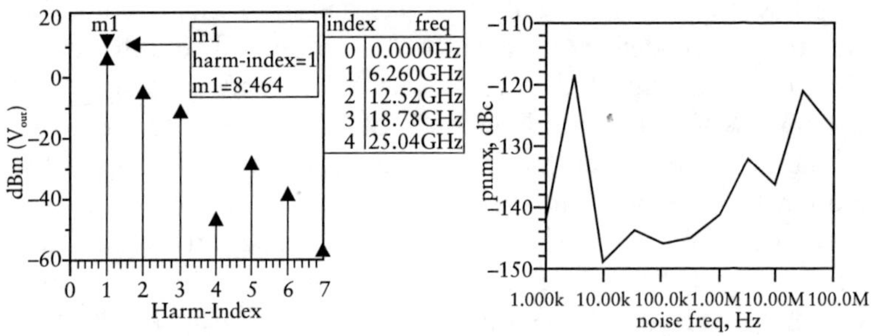

Fig. 6.34: Block schematic of the series feedback oscillator topology incorporating DR resonator

Further at S-band a stable DRO has been constructed and fabricated as shown in Fig. 6.35.

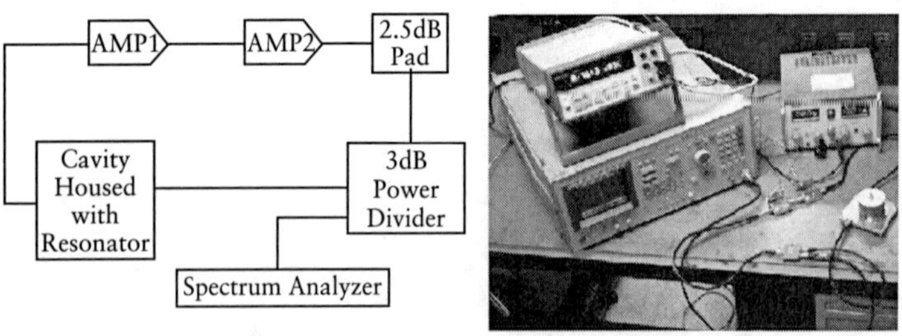

Fig. 6.35: Assembly of stable DRO

The output power measured is more than 1 mW and phase noise found out to be –113dBc/Hz at 10 kHz offset. The tuning range achieved ±50 MHz around 2 GHz.

6.6 Link Margin Analysis

The design of aerospace system is a complex procedure requiring trade-offs between many factors to achieve the best performance at an acceptable cost. Both excessive and low margin are undesirable which results in a lot of constraint on the electronic system. In aerospace applications, transmitter and receiver are the two systems responsible for commanding and ranging [2, 30]. The analog receiver block used has the potential to have lot of variations due to multiple sources and same is true for the transmitter block. Modern-day terrestrial and satellite communication systems demand optimization of each sub-system to achieve the desired specification. This section demonstrates the simulated performance of each block and the performance of the cascaded chain of both transmitter and receiver using SystemVue software. Simulated performance of each block provides an opportunity to predict and optimize the performance of each individual block. Further the link budget is calculated and the worst-case analysis is detailed in this section.

6.6.1 Concept of Link Analysis

Radio waves, which are used for wideband communications travel in line of sight at microwave frequencies and hence provide limited range of communication. There exists a requirement for a communication system which can cater to and extend the support to communicate across the globe. Since a repeater is needed to convey signals over long distances, satellites provide the ideal location to install repeaters.

Radio communication link can be established by various channels such as wires, coaxial cables, fibre optic cables, waveguide, atmosphere, empty space. Radio communication through wireless systems are preferred over the wired system for long range communication as the signal power in wired communication system attenuates exponentially with distance compared to space decay resulting in reduced signal strength. Wireless communication involves transfer of information in space using sound, optical, infrared, radio frequency techniques and most widely employed is the radio frequency communication as it offers wider bandwidth and can penetrate fog, foliage, dust, buildings and vehicles also. Connectivity for voice, video and data communications can be provided using wireless systems such as Wireless Local Area Networks (WLANs), Direct Broadcast Satellite (DBS) television service, paging systems, Global Positioning Satellite (GPS) service, Radio Frequency Identification (RFID) systems. Present-day systems employ higher frequencies to avoid spectrum crowding as well as offering wider bandwidth [1].

Wireless systems can be categorized as ground or satellite-based system and capable of providing multipoint-to-multipoint systems having full duplex capability for allowing simultaneous communication between the users. Satellite networks can provide wide area coverage and adopted in broadcasting internet, earth observation and military communications. In a satellite link delay from earth to satellite to earth is about 240 ms while in terrestrial link it will be far less due to placement of receiver on tall buildings or hilltops at certain distances. Very high bandwidths and very high data rates are achievable in a satellite-based communication system at higher frequency ranges at the expense of vulnerability associated with the weather conditions. Major advantage of satellite base link is minimal requirement of re-transmitters as are required in the terrestrial link. Transmission of signals over a satellite communication link requires Line-of-Sight (*LoS*) communication, but since theoretically three equidistant satellites in the geosynchronous orbit can effectively cover over 90 percent of the earth surface, the need for multiple re-transmissions is removed.

It becomes utmost importance to design the link budget with an optimum utilization of the space segment as well as feasibility of transmission of signals over long distances. Communication links is very critical and reliable communication link is strongly recommended for all communication systems from wireless to space based systems [31].

Optimal link availability and quality of performance depends on conduit factors and the content factors. The conduit factors include uplink-downlink path, effect on signal propagation, quality of earth station equipment, and the impact of the propagation medium in the frequency band of interest. The conduit factors deal mainly with the type of message transmitted and the devices involved in its transformation from one form to another as per suitability for transmission over a microwave medium.

6.6.2 Radio-frequency Communication Link

Received power is one of the most important factors for having a reliable radio communication link (Fig. 6.34). The received power by a radio antenna having circular aperture antenna of diameter D is given by Fris radio link formula as:

$$P_r = \frac{P_t G_t A_e}{4\pi R^2} = \frac{G_t G_r \lambda^2}{(4\pi R)^2} Pt = \eta\left(\frac{\pi.D}{\lambda}\right)^2 W \qquad \text{... (6.14)}$$

where P_t G_t = EIRP,

$\left(\frac{\lambda}{4\pi R}\right)^2 = FSL, A_e = \eta\frac{\pi D^2}{4} = \frac{7160}{f^2_{(MHz)}}, G_r = \frac{4\pi A_e}{\lambda^2}$ and in dB terms

$$P_r(dBW) = EIRP + G_r - L_p \qquad \text{... (6.15)}$$

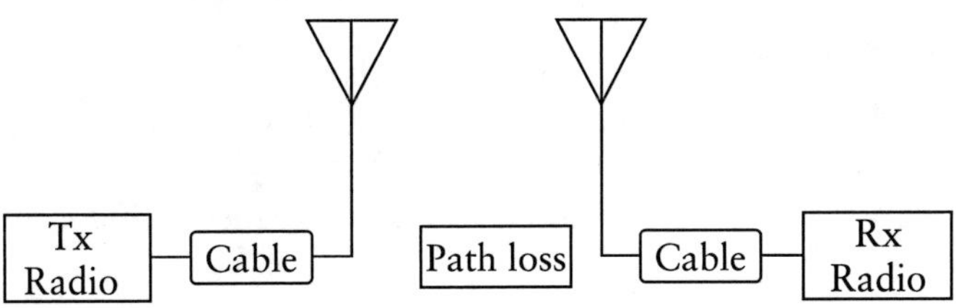

Fig. 6.36: Transmitter-receiver architecture for radio communication link

Alternatively, power received by an isotropic antenna by the known electric field strength E at the receiver is given as

$$P_r(dBm) = E\left(dB\frac{\mu V}{m}\right) - 20\log f_{Mhz} - 77.2 \qquad \text{... (6.16)}$$

The received power Pr is referred as carrier power (C) and is independent of the modulation scheme. Main functions of radio receiver are high gain, selectivity, down conversion, detection and transmitter isolation The calculation of link budget involves calculation of many parameters such as effective isotropic radiated power (EIRP), free space loss (FSL), carrier to noise ratio, bit error rate, various losses such as impedance and polarization mismatch of antennas, propagation effects and multipath effects to be taken into account apart from received power. Quality of a radio-frequency communication link is a function of various parameters such as receiver sensitivity, background noise level in the band, transmitted signal power level, transmitting and receiving antenna polarization and gain, dissipation/propagation losses and implementation losses [30]. Table 6.4 shows the comparison between the terrestrial and satellite based communication systems.

Table 6.4: Comparison between terrestrial and satellite communication

Parameters	Terrestrial Communication	Satellite Communication
Frequency	L-band	C-, Ku-, Ka-band
Atmospheric effects/ Rain attenuation	Negligible	High (changes with rain rate)
Material penetration	Better	Not good
Antenna Diameter	Larger	Smaller
Spectrum bandwidth	Lower	Higher
Beamwidth	Higher	Smaller
Link Loss	Lower	Higher (6 dB for twice the freq.)
Point-to-point services	Not good	Better
System Configuration	Simple	Complex
Delay time	Less	More
Antenna Configuration	Same at Rx/Tx end	Different antennas
Fade Margin	Less	More

Link budget needs the following information:

- Latitude and longitude of the uplink and downlink earth stations.
- Planned data or information rate.
- Modulation type (BPSK or QPSK).
- Forward error correction rate (1/2 or 3/4).
- Spread factor—if any (use only for spread spectrum systems).
- Uplink and downlink frequencies.
- Uplink and downlink antenna sizes.
- Uplink and downlink antenna efficiency.
- Uplink and downlink transmit and receive gains at frequency.
- Minimum digital signal strength (Eb/No) for desired Bit Error Rate (BER) performance.

6.6.3 Radio Propagation Losses

In free space waves travel in straight lines but in the presence of earth and its atmosphere the path of the wave gets altered. The bulk of the atmosphere extends to an altitude of 20 km. Radio propagation consists of mainly three electromagnetic waves, namely surface waves, sky waves and space waves. Wave propagation in high frequency region is mainly surface and sky whereas above high frequency the propagation is primarily space waves. During radio propagation electromagnetic waves suffer several atmospheric effects that results in power loss and these effects are: reflection and refraction, diffraction and scattering. These effects results in the large scale or small scale fading in accordance with the size of the object compared to the wavelength. In case of satellite-based system the main losses are categorized are: Free space losses, Atmospheric losses, Pointing losses and Inter symbol interferences.

Satellite link analysis is related to transmitted power, received power and accounting for the difference between these power levels. To analyse satellite link, basic transmission parameters such as antenna gain, beam width, free-space path loss and the link power equation are considered. The concept of system noise consisting of various parameters such as noise power, noise temperature, noise figure and figure of merit are used to account for channel characteristics. The carrier-to-noise ratio and related parameters are used to define communications link design and evaluating the overall performance.

6.6.4 Satellite Communication

The elements of satellite systems are earth segment and space segment. Earth segment comprises a network of transmit and receive earth stations whereas space segment comprises of a satellite [1-2]. The performance objectives of the satellite link is specified in terms of allowable (S/N) or BER for a given signal or as a minimum allowable carrier to noise power ratio C/N. Reliable link in satellite communication can be made by ensuring minimum signal-to-noise ratio (S/N) in the receiver baseband channel, optimizing transmitter power and RF bandwidth. S/N ratio is dependent on the carrier-to-noise ratio (C/N) of the RF or IF signal in the receiver, type of modulation and the RF/IF bandwidth in the receiver. The carrier power (C) and the noise power in the earth station receiver (N) can be represented as [6.17]:

$$\frac{C}{N} = \frac{P_r}{N} = \frac{P_t G_t G_r}{(4\pi R/\lambda)} \Big/ kT_0 B \qquad \text{... (6.17)}$$

The above value is specified in terms of percentage of outage time and at C-band it is taken as 0.01% of a year, at Ku band it is 0.1% of a year and for Ka-band 0.2% of a year.

$$\frac{C}{N} = \frac{P_t G_t}{kB}\left[\frac{\lambda}{4\pi R}\right]^2 \frac{G_r}{T_0}$$

So (C/N) α (G_r/T_0)

The ratio of G/T is known as figure of merit and is dependent on frequency and noise temperature. The noise generated in the receiver systems depends on the system noise temperature and the bandwidth used for communication. In case of downlink,

G/T (dB/K) = Receiver Antenna gain – 10 log (system noise temperature) (dB/K)

System noise temperature can be shown as (antenna is followed by line and receiver)

$$T_{sys} = T_{ant} + T_{cas} = T_{ant} + T_1 + \frac{T_2}{G_1} + \frac{T_3}{G_1G_2}$$

$$= \eta_{rad}\, T_A + (L_F - 1)\, 290 + T_{LNA} \times (L_F/G_R)$$

Noise can be lowered by employing better LNA with low noise figure while keeping antenna gain constant or else employing larger diameter antenna to be used to achieve the same performance. In case of uplink the noise contribution from sky and ground is prominent and changes with elevation angle. Interferences from other earth stations, terrestrial microwave links is considered as noise entering the receiver. Earth station G/T can be written as

$\frac{G}{T}(\deg/K) = \frac{8\cdot\pi\cdot k\cdot(Y-1)}{\lambda^2\cdot S\cdot K_1\cdot K_2}$ where K_1, K_2 are atmospheric and source correction factor $K_1 = K_g\cdot K_r\cdot K_d$

Main satellite parameters are Orbit, Orbit spacing, Orbital Inclination, Constellation size, Round trip delay, Polarization, Beam diameter, Transponder bandwidth, Antenna gain, Antenna diameter, Half power beamwidth, EIRP.

6.6.5 Link Budget Calculation

A link budget is simple addition and subtraction of gains and losses within an RF link. Link budget is calculated to estimate the practical performance of the system by taking into account the gains and losses introduced by different components. The computation of the link budget needs the following information:

- Latitude and longitude of the uplink and downlink earth stations.
- Modulation type.
- Uplink and downlink frequencies.
- Uplink and downlink antenna sizes.
- Uplink and downlink antenna efficiency.
- Uplink and downlink transmit and receive gains at frequency.

[LOSSES] = [FSL] + [RFL] + [AML] + [AA] +[PL] ...(6.18)

where,

[FSL] = free-space spreading loss, dB

[RFL] = receiver feeder loss, dB

[AML] = antenna misalignment loss, dB

[AA] = atmospheric absorption loss, dB

[PL] = polarization mismatch loss, dB

[PR]= [EIRP] – [LOSSES] ... (6.18a)

where,

[PR] = received power, dBW

[EIRP] = equivalent isotropic radiated power, dBW

6.7 System Level Simulation

There are several software tools that are currently available for the simulation of satellite transponders comprising a transmitter, receiver, diplexer, etc. and to establish a link margin for the entire communication system, link budget analysis. SystemVue simulation software is employed for the present case which enables system architects and algorithm developers to innovate the physical layer of wireless and aerospace communications systems and provides unique value to various implementers [31].

6.7.1 Design and Implementation

The basic building blocks for the implementation of the link budget model include the design of the transmitter and receiver [31-32]. Various parameters are later added to these models to compute the entire link budget analysis.

Transmitter: The input to the transmitter is an oscillator generating 0.1 V at a frequency of 27 kHz. This signal is fed to a phase modulator which generates a modulated waveform given by: [32]

$$s(t) = ACos\ (2pf_c t + k_p m(t)) \quad \text{... (6.19)}$$

with f_c (Carrier Frequency) = 350 MHz and k_p (Modulation Index) = 0.9 rad (51.566 deg)

The modulated signal is then passed through a frequency multiplier. The frequency translated signal is passed through a band pass filter to remove unnecessary sidebands. The signal is then amplified and given to a single pole double throw (SPDT)

RF switch to route the signal to Low-Power Amplification (LPA) and High-Power Amplification (HPA) paths. The signal is amplified in both the chains and two different output termed as low power and high power chain are taken out.

Transmitter Block Diagram: Each stage of transmitter is realised in the form of a block schematic as shown:

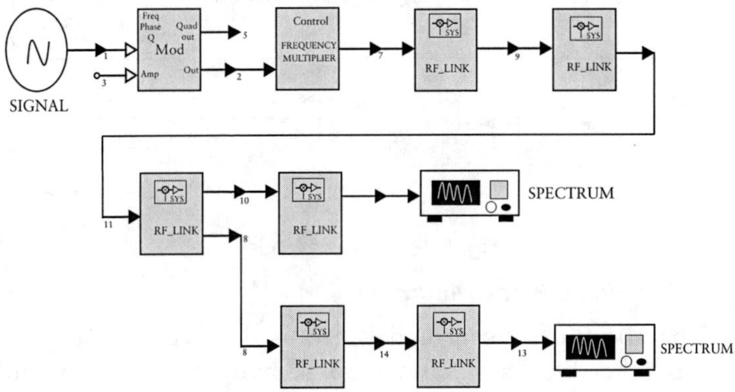

Fig. 6.37: Transmitter model

Receiver: The input to the receiver is a frequency modulated signal of the form [32]:

$$s(t) = A\text{Cos}(2\pi f_c t + 2\pi k_f \int_0^t m(t)dt) \quad \text{... (6.20)}$$

with f_c (Carrier Frequency) = 6400 MHz and k_f (Modulation Index) = 14.4 f_c

The receiver operation is considered at C-band (~6 GHz) and dynamic range ranges from –60 dBm to –115 dBm. The receiver is a super heterodyne type receiver that uses two-stage down conversion. The received signal is filtered by a band pass filter and then passed through a low noise amplifier to suppress the noise and amplify the signal. The signal is then down converted to get an intermediate frequency (IF). The down converted signal power is kept constant at 0dBm before the demodulation and fed to demodulator. The demodulated sinusoidal signal is amplified, filtered and passed through a power divider circuit to provide outputs for tele-command and ranging chains.

Receiver Block Diagram: Each stage of receiver is realised in the form of a block schematic as shown:

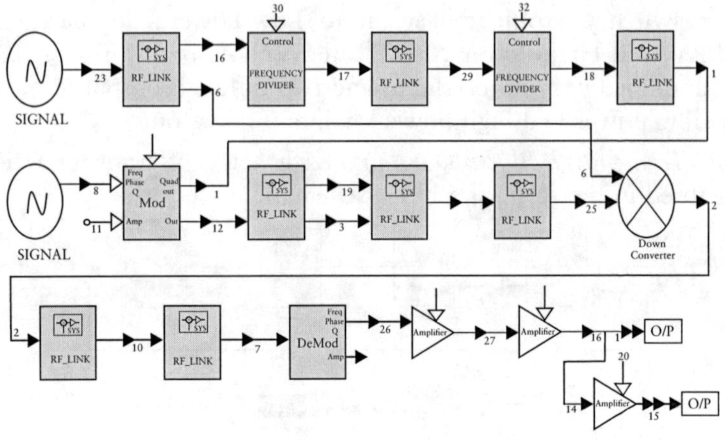

Fig. 6.38: Receiver model

6.7.2 Link Model Block Diagram

The above-described models are used together along with various other blocks to form the link budget model. The link budget model includes transmission of a frequency modulated signal from the earth station, reception at the satellite's receiver followed by demodulation of the signal, transmission of a phase modulated signal from the satellite transmitter and reception of the signal at the earth station receiver. All the losses during the transmission of signals are considered for the link budget model.

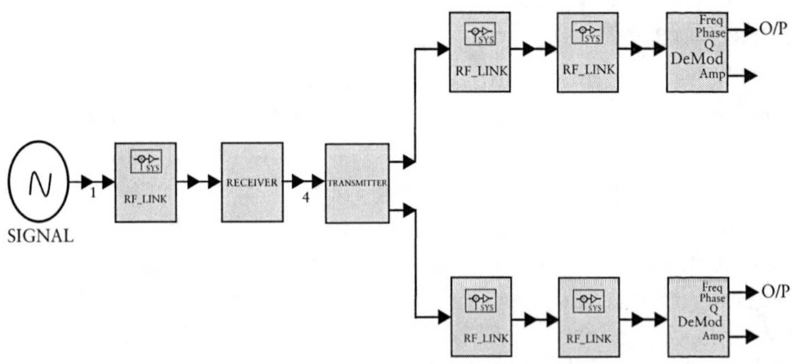

Fig. 6.39: Link budget model

Parameters and calculations involved in uplink path as shown in Table 6.5.

Table 6.5. Uplink parameters for the link budget model

Parameter	Lower Orbit (dBW)	Higher Orbit (dBW)
$P_T G_T$	85.0	80.0
Path Loss	–201.1	–201.1
G_R	–10.0	–3.0
Component induced on-board losses	–5.0	–5.0

For lower Orbit,

$P_R(dBW)$ = EIRP (dBW) + Path Loss (dBW) + G_R(dBW) + Component Induced Onboard Losses (dBW)

$= 85.0 - 201.1 - 10.0 - 5.0$

$= -131.1$

For higher orbit,

$P_R(dBW)$ = EIRP(dBW) + Path Loss(dBW) + G_R(dBW) + Component Induced Onboard Losses(dBW)

$= 80.0 - 201.1 - 3.0 - 5.0$

$= -129.1$

The demodulated output will be divided into two parts, the first output for command execution and another output provide the link signal.

Parameters and calculations involved in downlink path is shown in Table 6.6.

Table 6.6: Downlink parameters for the link budget model

Parameter	Transfer Orbit (dBW)	On Orbit (dBW)
PT	5.0	–4.0
Component induced on-board losses	–4.0	–4.0
GT	–10.0	16.0
EIRP	–9.0	5.0
Path loss	–197.43	–197.43

In case of lower orbit,

$$P_R(dBW) = \text{EIRP(dBW)} + \text{Path Loss(dBW)}$$
$$= -9.0 - 197.43 \qquad \Rightarrow = -206.43$$

In case of higher Orbit,

$$P_R(dBW) = \text{EIRP(dBW)} + \text{Path Loss(dBW)}$$
$$= 5.0 - 197.43 \qquad \Rightarrow = -192.43$$

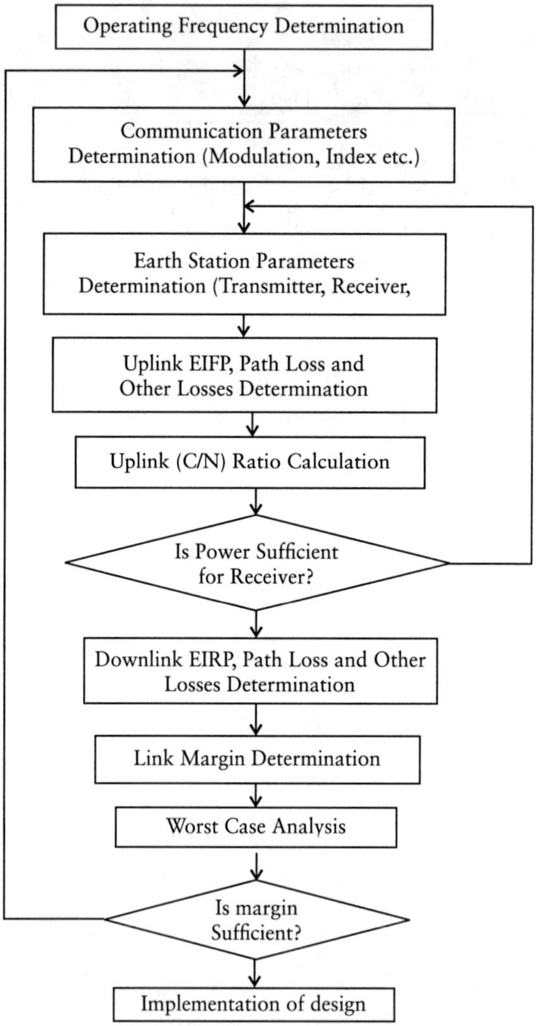

Fig. 6.40: Flow chart of link calculation

6.7.3 Results Analysis

The complete chain of communication link [33] is simulated and are shown from Fig. 6.41 to Fig. 6.46 and the results obtained are mentioned below:

(1) Transmitted signal from ground:

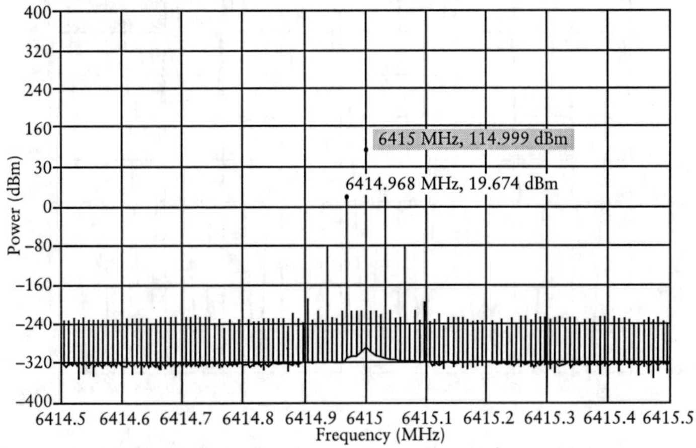

Fig. 6.41: Modulated output signal (as per simulation)

A frequency modulated signal is transmitted from the ground as shown in Fig. 6.41.

(2) Received signal at receiver end:

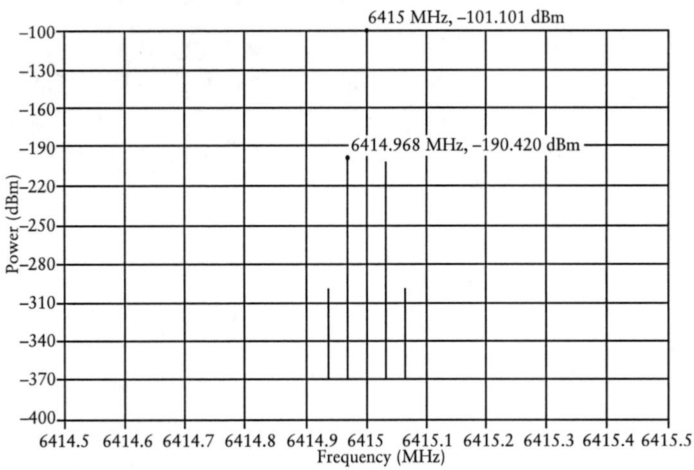

Fig. 6.42: Signal received at the receiver end (as per simulation)

The transmitted signal from ground through a communication channel and the attenuated received power is shown in Fig. 6.42.

(3) **Demodulated signal at receiver:**

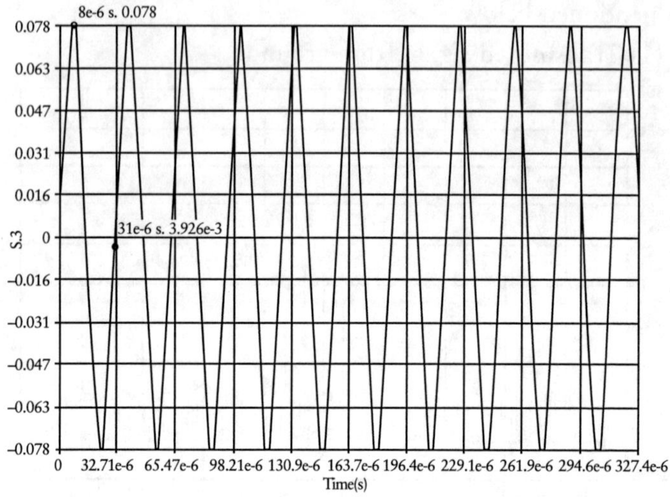

Fig. 6.43: Demodulated output signal at the receiver (as per simulation)

The demodulated signal (Fig. 6.43) from receiver is given as modulating signal to transmitter [34].

(4) **Transmitted downlink power:** A phase modulated signal is transmitted to the ground with a RF signal power of about 5 dBW as shown in Fig. 6.44.

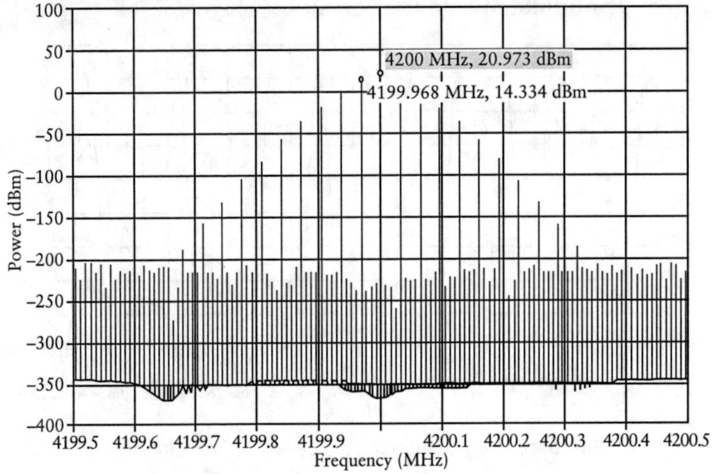

Fig. 6.44: Transmitter output (as per simulation)

(5) **Downlink received power:** The transmitted signal from transmitter reaches the ground through a communication channel and the attenuated received power is shown in Fig. 6.45.

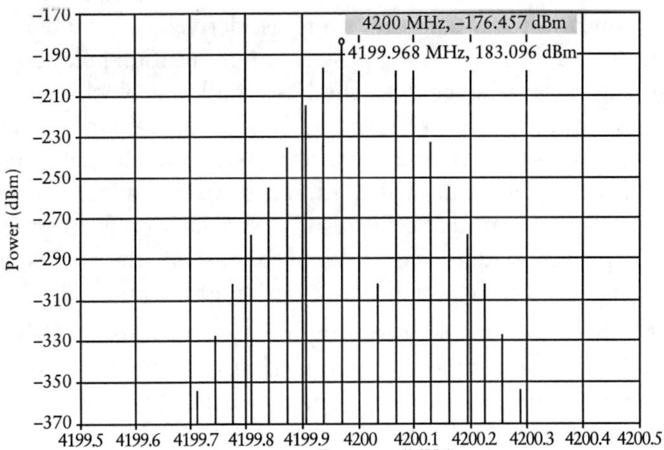

Fig. 6.45: Signal received at the ground (as per simulation)

(6) **Demodulated signal at ground:**

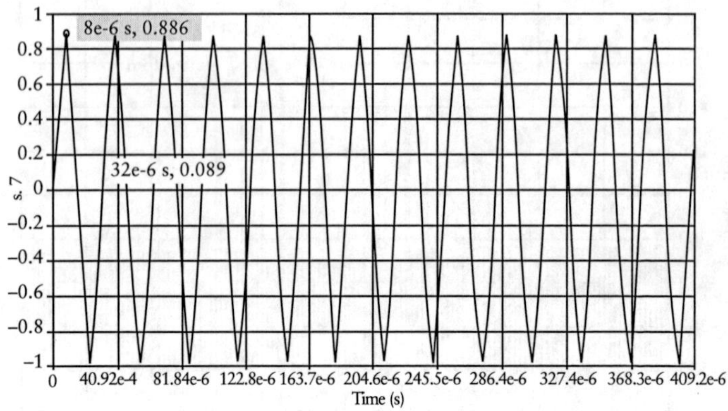

Fig. 6.46: Demodulated output (as per simulation)

The received phase modulated signal [35] can be recovered at ground as shown in the Fig. 6.46. The recovery of the signal depends on the sufficient margin in which the transmitted signal from ground is compared with the received signal.

The analysis of link budget is very important for any communication system and same is derived, explained and implemented. In order to establish a communication link between earth station and spacecraft, the basic building blocks, namely earth station, on-board receiver, on-board transmitter, antennas, etc. are simulated, explained and demonstrated. Here, a ranging signal transmitted from earth station is recovered at earth station and same is compared and the result is satisfactory. Furthermore, the sample link budget was outlined to illustrate the process and the steps involved in satellite link design are detailed.

Methodology of incorporating mixer instead of SRD-based multiplier is easily implementable and has potential of realization with shorter turnaround time. The implementation of the same needs minimum modification in the existing topologies and needs less realizable time due to avoidance and dependence on multiple tuning. Further the concept is shown to be extendable for Ku-band also which needs incorporation of the active multiplier for low order multiplication. Selection of the components are carried out as per availability as per industry it can be easily productionized. The added advantage of the concept is removal of temperature variability associated with SRD-based approach and due to riddance of tunability and incorporation of MMIC, turnaround time is less. Future system to be developed catering for high power, high bandwidth Ka-band satellites, ground systems for point-to-point connectivity and it needs paradigm shift in the system configuration having incorporation of re-configurability embedded in it. Ka-band system to accommodate wider-bandwidth modulation schemes and higher data rates, which include achieving mm-wave power levels and RF power added efficiency (PAE), frequency and phase stability, and phase noise. Similarly, other aspects such as amplitude and phase equalization across GHz-wide operating channels, digital modulation and demodulation performances to be critically examined along with the concept of RF-CMOS so as to realize active and passive circuits on the single chip such as mixer, RF and low noise amplifier.

6.8 Various System Level Implementation Approaches

As the multiplier-based topologies employing SRD are prone to variations resulting in many bottlenecks in high end system. The main challenges such as extensive tuning and tweaking, EMI problems, components obsolescence, etc. The main limitation of diode-based multiplier is the inherent non-linearities resulting in batch-to-batch variation in the device. The impulse generator circuit consists of the inductor in series with the shunt capacitor followed by resonator circuit to resonate at the output frequency. The realization of these circuits is carried out employing tunable capacitors which caters for the proper matching and proper operation of the multiplier chain. Sensistor for self-bias is also incorporated with capacitor to remove parametric instabilities associated with this circuit and slight change in power results in degraded performance. Keeping into consideration of the above tunablity option, the circuit needs extensive optimization at card and package level to achieve the desired results. Table 6.7 provides the various critical aspects associated with the generic approach [36].

Table 6.7: Critical aspects

Component	Shortcomings	Time Consumption	Mitigation Techniques
TCXO	Spurious components	<10%	Stable source with low spurious components
Multipliers	Tuning	~50%	Proposed approach
Problem diagnosis	Fab related/ component	~10%	Listing of frequent occurring problems

This necessitates to change the topology to mixer based, MMIC or RFIC based as well as concept of digitatization after second IF using ADC. These aspects are briefly discussed in the subsequent sections.

6.8.1 Transponder Description

Generic telemetry and telecommand transponder (Fig. 6.46) consisting of transmitter and receiver which are optimized separately consisting of multiple cards. The main challenge is to integrate all the circuits and having the optimized performance.

Transmitter PCB in MIC having different material such as RT duroid, FR4 substrate for optimized performance which includes components such as FETs, quadrature hybrid, SRD, capacitor and resistor. Standard transmitter consists of multiple sections [37] such as.

- Temperature controlled crystal oscillator (reference source)
- Multiplier.
- Modulator.
- Power Amplifiers.
- Power Supply and Interface circuitry.

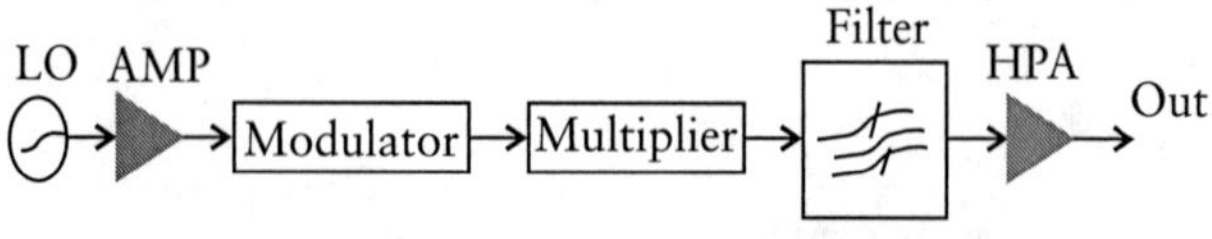

Fig. 6.47: Standard transmitter topology

Main parameters in transmitter are:

- Output power (various mode)
- Harmonics and spurious
- Frequency stability of source
- Modulation frequency response
- Modulation indices
- Modulation format
- Group delay

Receiver in double super heterodyne configuration is implemented with multiple components such as: LNA, SAW filter, multipliers, amplifiers, attenuator and local oscillator block [Fig. 6.48]. Receiver PCB having different material such as RT duroid, TFG, Glass epoxy, Alumina substrate realization carried out for high frequency circuits such as LNA, RF amplifier, mixer.

The radio-receiver provides various functions such as: high gain, selectivity, down-conversion, detection, isolation from the transmitter. The selectivity of the receiver is met with the narrow band pass filter with sharp cut-off which is generally met with the waveguide filter or saw filter.

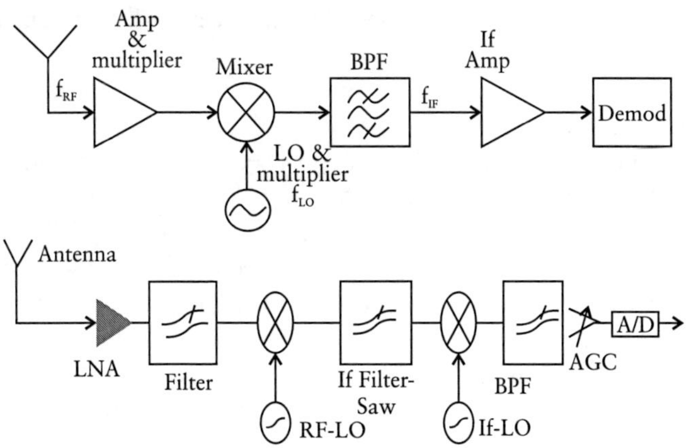

Fig. 6.48: Standard super-heterodyne receiver topologies

Main parameters in receiver are:

- Noise figure
- Selectivity

- Isolation
- Dynamic range
- Modulation scheme
- Bandwidth
- S/N threshold
- Modulation indices
- Modulation format
- Group delay

Present trend is to have direct conversion so as to remove circuit complexity, increased flexibility, weight and power consumption, but this approach was earlier limited due to the unavailability of ADC/DAC at high frequency. Presently, direct conversion to and from Ka-band frequencies is now possible with advances in wide bandwidth data converters at Ka-band having analog bandwidth of 20 GHz.

Frequency Synthesizer-based Approach

Frequency synthesizer are used to generate LO signal in the transmitter and receiver which is based on the PLL control systems. Phased locked loop (PLL) is employed in various communication applications such as clock generation, frequency synthesizer, clock recovery, radio receiver, etc. Phase Locked Loops are mainly utilized for clock synthesis, synchronization, skew and jitter reduction. In order to reduce the overall power dissipation of the circuit, phase lock loop architecture is used for designing frequency multipliers. PLL-based approach eliminates requirement of high order multiplier which is thus easier to implement using FET-based topologies. Phase locked loop is the indirect method for generating the output signal which is related with the input signal. VCO is incorporated in the feedback control loop for controlling the phase in the closed system. The technique for frequency synthesizer is associated with the integer and fractional-N techniques for generating the desired output frequency [38]. Main parameters associated with the PLL synthesizer are-step size, frequency range, switching speed, spur, and phase noise. Switching speed is the function of loop bandwidth (step size) which controls the switching speed [39]. Basic architecture of PLL consists of: Phase frequency detector (PFD), low pass filter (LPF), Voltage controlled oscillator (VCO) [Fig. 6.49]

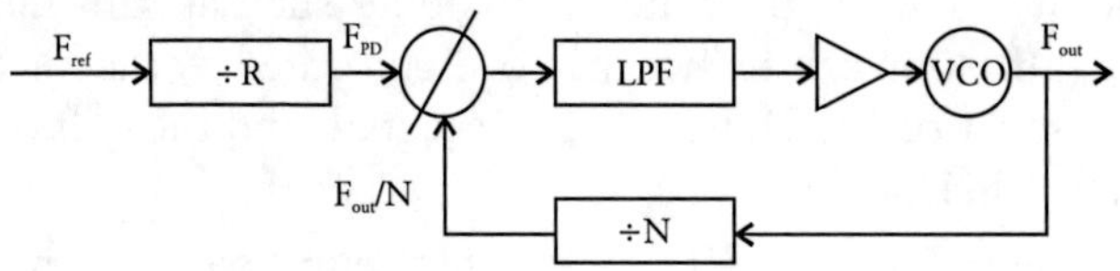

Fig. 6.49: PLL schematic (F_{out}=(N F_{ref})/R)

The phase noise impact many critical parameters in the synthesized oscillator and accordingly selection to be carried out. More detailing of the integer-N PLL is shown in Fig. 6.50.

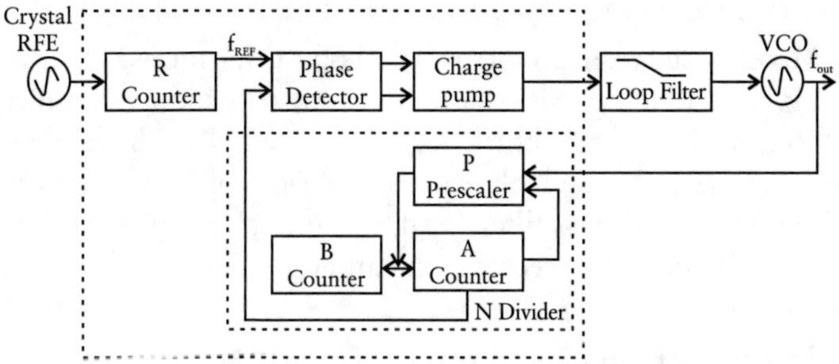

Fig. 6.50: Integer-N system

Frequency synthesizers are selected based on low phase noise, low spurious, frequency agility and lower frequency hopping time. A generic trans-receiver architecture employing PLL is shown in Fig. 6.51.

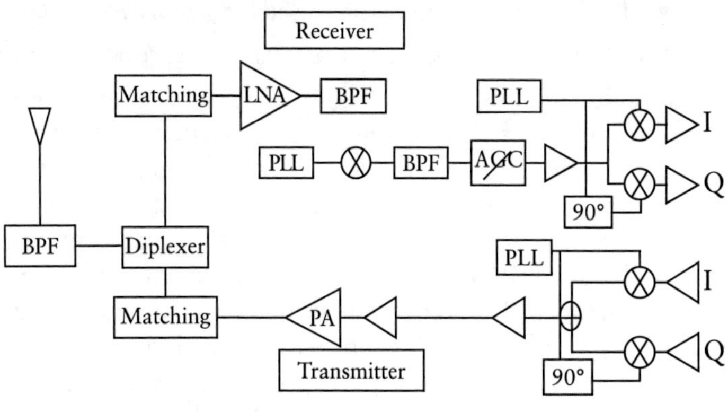

Fig. 6.51: Trans-receiver architecture

Frequency synthesizers generally suffer from quantization noise, which causes them to exhibit poor phase-noise performance outside their loop bandwidth. Techniques such as self-injection locking (SIL) loop can reduce the noise generated by the injection-locked oscillator (ILO), the reference signal, and the delta-sigma modulator [41].

Frequency multipliers are often employed with local oscillators (LO) operating at lower frequencies for microwave and millimeter-wave (mm-wave) signal generation and having added advantage of low cost [44]. The pair of oscillator-multiplier pair alleviates many design constraints of the oscillator and related circuitry such as in phase locked loops (PLL) [42].

Mixer-based transponder is based on the harmonic generation in mixer replacing SRD as mixer also produces multiple harmonics. The main bottleneck in mixer is to have precise frequency selection which can be carried out by employing cavity-based band pass filter for harmonic selection [Chapter 4]. It can be direct or indirect which can be implemented in analog or digital domain.

Transmitter Configuration

- Mixer-based multiplier eliminating SRD-based topology.
- Internally matched MMIC amplifiers and GaN-based topology for high power output.

Receiver Configuration

- FPGA-based demodulator is used for signal demodulation.
- PLL-based synthesized source is used for carrier generation.

The major advantages of this topology are: elimination of SRD multiplier cards with mixer-based frequency multiplication (no tuning required), synthesized-based version, active multiplier for frequency multiplication (Ku-band implementation), MMIC amplifier (single assembly) for low power mode and internally matched FET for HPA mode. The same concept can be extended for the Ka-band (30 GHz) where frequency agility synthesizer employing agile and low phase noise local oscillators along with direct frequency synthesizer (DDS) up-conversion technique to generate linear FM signal (Fig. 6.52). This concept to be implemented in place of PLL which has limitation of low phase noise and fast switching speeds whereas DDS can generate linear and agile frequency chirps.

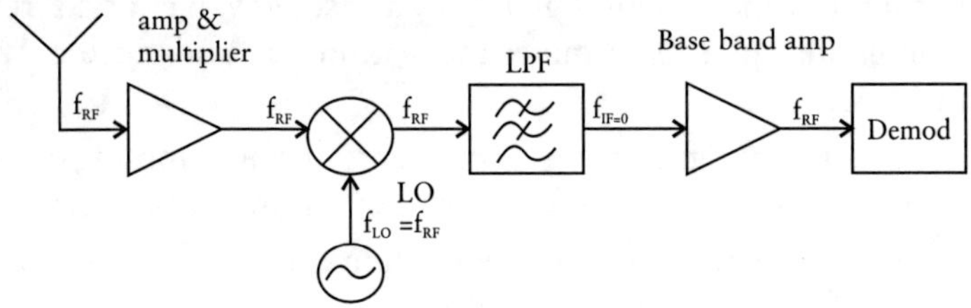

Fig. 6.52: Direct conversion receiver

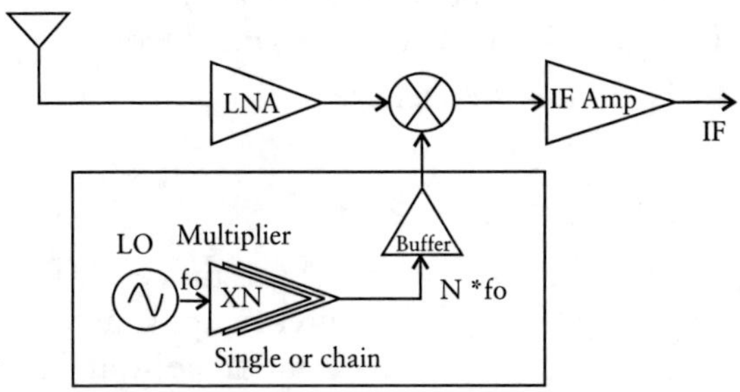

Fig. 6.53: Multiplier approach implementation [42]

The main advantage of the PLL is the reduction of the spurious levels and lower level of complexity to the direct analog architecture. The PLL synthesizer characteristics can be improved by employing fractional-N technique. The main disadvantage of the PLL techniques compared to standard approach is the slower tuning and higher phase noise.

Millimetric wave frequency synthesizers are PLL with VCO, direct digital frequency synthesizer (DDS) driving PLL, DDS frequency up-conversion in PLL loop.

6.8.2 Digitization of RF Signals

The present trend of employing higher frequency ADC for sampling eases lot of constraint related with tuning and optimization of the RF system apart from reducing the size and number of components. ADC samples an analogue waveform at uniform time intervals and assigns a digital value to each sample using mainly two processes: sampling and quantization. Sampling is simply the reduction of a continuous time signal to a discrete time signal and sampling frequency is the average number of received samples in one second. The sampling process represents a continuous time domain signal with measured values at discrete time intervals and it determines the maximum bandwidth of the sampled signals.

In order to avoid aliasing components during sampling of the high-frequency signals the following conditions to be met [43]:

1. Sampled signal bandwidth should not exceed one half of the sampling rate (fs/2)
2. The maximum and minimum frequencies of the sampled signal should be within the same Nyquist zone
3. The sampling frequency

$$f_s = \frac{5f_{rf}}{2N_Z - 1} \qquad \text{... (6.21)}$$

where N_z is the Nyquist zone number, f_{rf} is the center frequency of an arbitrary RF input signal. The three major sources of noise in the system are: thermal noise, quantization noise and phase noise due to clock generator. The noise power can be represented as:

$$P_n = P_{n,\ phase\text{-}noise} + P_{n,\ quantization\text{-}noise} + P_{n,\ thermal\text{-}noise}$$

$$P_{n,quantization} = \frac{P_{input,ADC}}{SNR}$$

$$P_{n,phase-noise} = \frac{P_{input,ADC}}{SNR_{phase-noise}}$$

$$P_{n,thermal} = k_B.T_0.B.NF.G$$

$$SNR_{phase-noise} = -\frac{1}{(2\pi f_{RF} t_{jRMS})^2}$$

$$t_{jRMS} = \frac{1}{\omega_s}\sqrt{\int_0^{\infty} 2.L(f)df}$$

where L(f) is the ratio between the phase noise power level and the signal power level at a frequency offset. At low power levels dominance of thermal noise compared to other two whereas at higher power levels phase noise is prominent. Quantization noise is too low compared to thermal and phase noise. The major advantage of digital down conversion is the improvement in the SNR by 10 log (BW_{in}/BW_{out}).

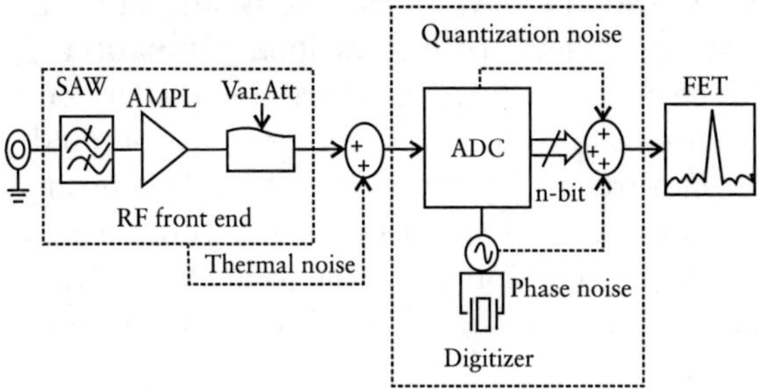

Fig. 6.54: Concept of digitization of RF signals

Digital receiver after first or second IF stage for LO frequency also require multiplication for higher frequency implementation. In case of heterodyne receivers sampling approach (under-sampling) can replace mixer based implementation.

Fifth generation (5G) transmitters with wideband modulation signal needs linearization bandwidth constraint due to maximum clock frequency of the digital circuit. Under-sampling scheme using a single low-speed ADC (Fig. 6.55] can minimize the complexity and power consumption that typically scales with the linearization bandwidth [42].

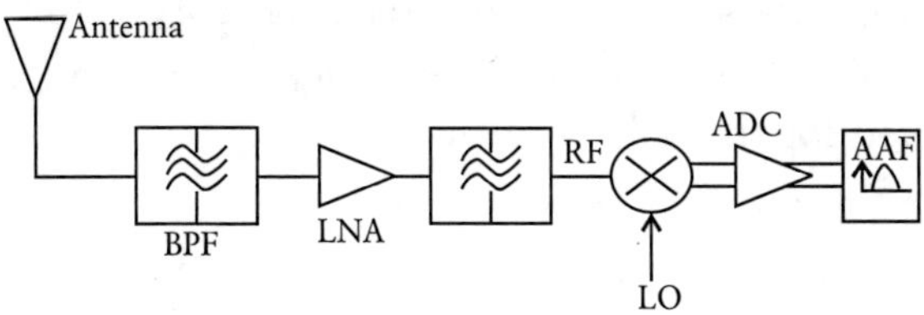

Fig. 6.55: Digital filter concept

Direct-Conversion Receiver

RFIC-based wireless systems are presently having direct-conversion receiver architecture having 0 IF (DC) wherein LO frequency exactly equals to RF frequency. This approach yields considerable advantages compared to the standard heterodyne architecture such as image signal itself is the desired signal eliminating the need of filter, IF stages, IF-VCO, PLL, etc. Another aspect related with the direct conversion to digital domain with the incorporation of ADC and digitally performing channel selection with DSP employing FPGA but needs high dynamic range and selectivity [45]. This results attractive features at RF such as higher level of integration, elimination of off-chip filters, reduce parasitic, lower cost and low power consumption. The main limitations of this approach are: limited dynamic range, second-order distortion, dc-offset, I/Q mismatch. The problem related with DC offsets is more prominent on silicon based implementation approach due to low-resistivity substrate coupling; as strong LO signal can leak and can interfere with the RF signal.

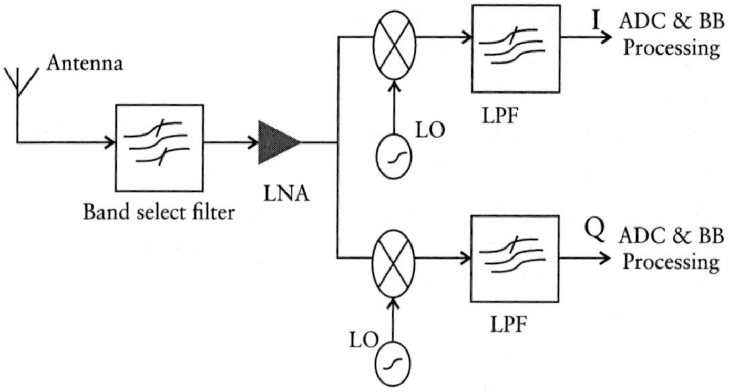

Fig. 6.56: Schematic of direct conversion receiver (DCR)

In case of digitization of the signal, the SNR can be represented for the N-bit digitizer as:

$$\text{SNR} = 1.76\text{ dB} + 6.02 \times N$$

The noise is mostly quantization noise but figure of merit can be represented by ENOB which is the summation of measured SNR and distortion power ratio and can be represented as:

$$\text{ENOB} = (\text{SINAD}-1.76)/6.02$$

Frequency multipliers are frequently implemented alongside VCOs, PLLs, amplifiers and mixers in a variety of system applications such as transceivers, frequency synthesizers, down converters, commercial digital broadcast satellite transceivers

(DBS), broadband wireless systems, automotive radar and phased array applications.

CMOS Technology

Present trend is to realize CMOS-based multipliers having the potential of complete circuit realization on a single compact chip but these are having higher losses. Many of the performance improvements and development directions of frequency multipliers are driven by advances in semiconductor technologies such as InP, GaAs metamorphic HEMT (mHEMT), SiGe BiCMOS and AlGaN/GaN (GaN). Mainly MOS models are: BSIM, Berkleley, HSICE. Present-day technologies are the amalgation of the various techniques on CMOS such as BiCMOS, SiGe-BiCMOS. Due to decrease in the feature size in CMOS, speed of CMOS devices increased as well as cut off frequency (f_T). This results in the design of logic and analog circuits at higher frequencies. The figure of merit (FOM) for the technology selection is based on the f_T, f_{max}, g_m, g_0, 1/f, and noise characteristics. In CMOS technology, scaling of gate length results in the higher transconductance but compromises on thc breakdown voltage and short channel effects. So present trend of implementing higher frequencies ADC or DAC (ASIC/ FPGA) is limited due to sampling rate and resolution directly at higher frequencies apart from slower rate of computation and consumption of more power due to fast charging and discharging. Direct digitations also calls for broadband circulator and tight selectivity requirements which is challenge at present.

RF-SOI based technology is gaining popularity due to various advantages such as:

- n-FET in SOI is having lower parasitic than bulk CMOS.
- Power output (10x), f_{max} (>25%), efficiency (2-5 times) can be achieved using SOI.
- Transistors, substrate and back end provide all having improved performance.

Recently, linear transmission line (NLTL)-based frequency multipliers which are having the advantage of high output power and broadband operation is demonstrated by developing a 20-GHz frequency doubler using 65-nm CMOS process and 100-

GHz tripler using 130 nm technology is demonstrated. Quantum receivers are also being implemented covering full range from 1-100 GHz such as Ryderg atom.

Table 6.8: Various technologies and its comparison

Parameter	GaAs	SiGe	SiCMOS
Noise	Excellent	Excellent	Good
Cut-off frequency	Excellent	Excellent	Very good
Isolation	Excellent	Excellent	Good
Integration level	Good	Good	Excellent
Power Consumption	Good	Excellent	Good

Other aspects related with the signal generation at high frequencies is the fitter noise which is an important parameter for high order QAM and millimetric wave [39]. Impact of the oscillator on MIMO technology relates with the phase noise which degrades channel aging.

This chapter shows that system level simulation can provide efficient and accurate results in RF communication systems setup. The transmitter configuration at S-band for TT&C applications employs frequency synthesizer, I/Q modulators, power amplifiers and mixer as shown by M. Kameche [40]. The RF signal is directly converted to the baseband and amplified which provide better performance due to elimination of cross talk and various parasitics.

References

1. D.M. Pozar, *Microwave Engineering*, John Wiley & Sons, 2010.
2. Dennis Roddy, *Satellite Communications*, McGraw Hill Education, 2005.
3. C. Quendo, E. Rius and C. Person, "Narrow Bandpass Filters Using Dual-Behavior Resonators," *IEEE Trans MTT*, Vol. 51, pp. 734-42, March 2003.
4. R.V. Snyder, "Practical aspects of microwave filter designs," *IEEE Microwave Magazine*, April 2007, pp. 42-54.
5. J.A. Ruiz-Cruz, C. Wang and K.A. Zaki, "Advances in Microwave Filter Design Techniques," *Microwave Journal*, Nov. 2008, pp. 26-44.
6. E. Ruis, Y. Clavet, C. Quendo, A. Manchec, O. Bosch, C. Person, J. Favennec, J. Cayrou and J.L. Cazaux, "Comparison Between Ku-band

Classical and Cross-coupled Microstrip Hairpin Filters," European Microwave Conference, Oct. 2005.

7. A. Manchec, E. Rius, C. Quendo, C. Person, J. Favennec, P. Mironi, J.L. Cazau, "Ku-band diplexer based on Dual Behaviour Resonator Filter," 2005 *IEEE MTT-S International Microwave Symposium Digest*, pp. 525-28.
8. E. Rius, C. Quendo et al., "High Rejection C-band Filter for a Spatial application," 33rd European Conference, Munich 2003, pp. 1055-58.
9. Daniel G. Swansen and W.J.R. Hoefer, *Microwave Circuit Modeling Using Electro-magnetic Field Simulation,* Artech House, 2003.
10. Kamaljeet Singh, K. Nagachenchaiah, D. Bhatnagar and S. Pal, "Notch Implemented Dual Behavior Resonator (DBR) Filter and Diplexer at Ku-Band," *Microwave Journal*, 2008.
11. C. Quendo, E. Rius and C. Person, "Narrow bandpass filters using Dual-behavior resonators based on stepped-impedance stubs and different length stubs," *IEEE Trans MTT*, No. 3, Vol. 52, pp. 1034-44, March 2004.
12. Gullimero Gonzalea, *Microwave Transistor Amplifiers Analysis and Design*, Prentice Hall, New Jersey 1984.
13. Y. Park, C.H. Lee, J.D. Creesler and J. Laskar, "The analysis of UWB SiGe HBT LNA for its noise, linearity, and minimum group delay variation," *IEEE-MT Trans*, Vol. 54, April 2006, pp. 1687-97.
14. Bagher Afshar, A.M. Nikngad, "X/Ku-band CMOS LNA design techniques," IEEE-CICC, 2006, pp. 389-92.
15. Mike Golio, *RF and Microwave Semiconductor Device Handbook*, CRC Press, 2003.
16. A. Grebeemnikov, Narendra Kumar, S. Binbaga, *Broadband RF & Microwave Amplifier*, CRC Press.
17. Kamaljeet Singh, *Interdigital Based Capacitor,* Atlantic Publishers and Distributors (P) Ltd., 2020.
18. Domine Leenaerts, Johan vander Tang, and Cicero S. Vaucher, *Circuit Design for R.F. transreceivers*, Kluwer Academic Publishers, 2001.
19. B. Kim, D. Derickson and C. Sun, "A High Power, High Efficiency Amplifier using GaN HEMT," Asia-Pacific Microwave Conference 2007, pp. 1317-20.
20. J. Coonrod, "Selecting circuit materials for Microwave Power Amplifier," *Microwave Journal*, 2015.
21. M. Leszczynski, P. Prystawko et al., "Comparison of Si, Sapphire, SiC and GaN substrates for HEMT epitaxy," Abstracts from The Electrochemical Society, 2012, Honolulu.

22. V.F. Fusco and J.A.C. Stewart, "Design and Synthesis of Patch Microwave Couplers," 16th European Microwave Conference, 1986, pp. 401-06.
23. Jui-Chich Chiu, Jhi-Minghin and Yeng-Her Wang, "A novel planar three way power divider," *IEEE Microwave and Wireless Components Letters*, Vol. 16, No. 8, Aug. 2006, pp. 449-51.
24. M.D. Abouzahra and K.C. Gupta, "Multiport power divider-combiner circuits using circular-sector shaped planar components," *IEEE Transactions on Microwave Theory and Techniques*, Vol. 36, No. 12, Dec. 1988, pp. 1747-51.
25. Lin Li and Ke Wu, "Integrated Planar Spatial Power Combiner," *IEEE Transactions on Microwave Theory and Techniques*, Vol. 54, No. 4, April 2006, pp. 1470-76.
26. Hao Guo, Chun-Qing Chen, Hao-Quan Wang and Ming-Li Hao, "A 2-4 GHz Octave Bandwidth GaN HEMT Power Amplifier with High Efficiency," *Progress In Electromagnetics Research Letters*, Vol. 63, 7-14, 2016.
27. Chen, Chang and Ching, "A unified design of dielectric resonator oscillator for telecommunication system", *IEEE MTT-S Digest*, 1986, pp. 593-96.
28. R.R. Boneti and A.E. Atia, "Analysis of microstrip circuits coupled to dielectric resonators," *IEEE MTT*, Vol. 29, No. 12, Dec. 1981, pp. 1333-36.
29. A.P.S. Khanna and Y. Garanault, "Determination of loaded, unloaded, and external Quality Factor coupled to a microstrip line," *IEEE MTT*, Vol. 31, No. 3, 1983, pp. 261-64.
30. Fiedzuszko and Holme, "Dielectric Resonatore-Raise your high-Q", *IEEE Microwave magazine* 2001, pp. 51-60.
31. S.K. Sharama, S. Chatzintos and B. Otersten, "In-line interference mitigation techniques for spectral coexistence of GEO and NGEO satellites," *International Journal of Satellite Communications,* Vol. 34, No. 1, 2016, pp. 11-39.
32. Aderemi A. Atayero, Matthew K. Luka and Adeyemi A. Alatishe, Satellite Link Design: *A Tutorial, International Journal of Electrical & Computer Sciences IJECS-IJENS,* Vol. 11, No. 04, 2011, pp. 1-6.
33. Haykin, S. and Moher, M. (2012). *An Introduction to Analog and Digital Communications*. Hoboken: Wiley Textbooks.
34. R. Ashiya, "A regional satellite system for mobile communications," *ICPWC,* 94, pp. 142-46, 1994.
35. C. Haslett, *Essentials of Radio Wave Propagation*, Cambridge Wireless Essentials Series, 2008.

36. Timothy Pratt et al., *Satellite Communications*, 2003.

37. K. Singh, R. Selvi, A.V. Nirmal, S.V. Sharma, "RF sub-systems: Productionization perspective and challenges," ESSRI Conference, ISRO, 2016.

38. Suman R. Valke, T.S. Narayanan, K.S. Sudhira, M. Neelavathy, D. John and K.N.S. Rao, "A New Miniaturized Version of C-band TT&C Transmitter for INSAT," ISAC Library, pp. 32-38.

39. Pasternack, "The growing importance of oscillators with 5G,", *Microwave Journal*, pp. 52-62, Vol. 63, No. 8.

40. Mohamed Kameche, "Simulating and Designing an RF transmitter for small satellites," *High Frequency Electronics Magazine*, May 2007.

41. Alexander Chenakin, *Frequency Synthesizers: Concept to Product*, Artech House 2011.

42. *IEEE Transactions on Microwave Theory and Techniques*, online 2020.

43. Kelvin Yuk, "Advances in Active Microwave Frequency Multipliers," EP, August 2011.

44. Uros Mavric, *Digitizer provides Direct Sampling of RF Signals, Microwaves & RF*, April 2006.

45. S.A. Maas, *Non-linear Microwave Circuits,* 2nd Ed., Artech House, 2003.

46. R. Svitek and S. Raman, "DC offsets in Direct-Conversion Receivers: Characterization and Implications," *IEEE Microwave Magazine*, Sept 2005, pp. 76-86.

Annexures

Annexure 1

Measurement Units in RF Domain

dB - It is not a physical quantity but ratio of two physical quantities.

dB = Ratio of two power levels

$= 10 \log_{10} (P_2/P_1)$

In case of voltage then,

dB = $20\log_{10} (V_2/V_1)$ as $P = V_2/R$

(in case of equal impedances)

dBW = Expression of power with respect to reference power level of 1W

dBW = $10 \log_{10} (P_2/1W)$

dBm - Expression of power with respect to reference power level of 1mW

dBm = $10 \log_{10} (P_2/1mW)$

dBμV = Voltage ratio in dB referred to reference voltage of 1μV

Table A1.1: Ratio Vs Decibel

Ratio	dB
1	0 dB
2	3 dB
3	5 dB
4	6 dB
6	8 dB
10	10 dB
100	20 dB
10^3	30 dB

Conversions in dB Scales

dBW = –30 + dBm (1W = 10^3 mW)

= –60 + dBμ W (1W = 10 μW)

dBW = 10 log (P_2W/10^3mW)

= –30 + 10 $\log_{10}$ (P_2/1W)

dBV = –60 + dBmV

= –120 + dBmV

dBV = 20 $\log_{10}$(1V/10^3 mV)

= 20 [$\log_{10}$ (10^{-3}) + $\log_{10}$(v/m]

= –60 $\log_{10}$ (1V/1mV)

= –60 + dBmV

1W = 0 dB = 30 dBm

Table A1.2: Correlation with various parameters

Parameter	Resultant	Co-relation
dB+dB	dB	Product of two numbers
dB-dB	dB	Quotient /ratio
dBm+dBm	invalid	Power levels
dBm-dBm	dB	Comparing power levels
dBm+dB	dBm	Amplification
dBm-dB	dBm	Attenuation

Concept of various parameters

P_{in}- input power

P_{out}-output power

P_{ref}-reflected power

Insertion Loss= P_{in}/P_{out}; Power Gain = P_{out}/P_{in}; Return Loss = P_{in}/P_{ref}

Main Identities

I. Product of two pure numbers is equivalent to their sum when it is expressed in dB.

A = 1 (0 dB)

B = 2 (3 dB)

A × B = 2 (0 + 3) = 3 dB

II. The division of two pure numbers is equivalent to their difference when their values are expressed in dB.

X = 4 (6dB), X/Y = 4/2 = 2 (6–3) dB = 3 dB

Y = (3 dB)

Fractional Numbers

Ratio	dB
0.1 = 1/10	0 dB/10 dB

So 0.1 = 0 dB – 10 dB

0.1 = – 10 dB

Negative numbers cannot be represented in dB.

Converting dB to pure numbers and vice-versa

I. Divide the given dB value by 10

II. Use anti log of (dB/10) to obtain number

Example: 5 dB

Step 1 ÷ 5/10 = 0.5

Step 2 ÷ anti log (0.5) = 3

Converting 2W into dBm

2W = 2×10^3 mW

Now dBm = 10 log (2×10^3mW/1 mW)

= $10 \log_{10} (2 \times 10^3)$

$= 10 \log_{10} (2) + 10 \log_{10} (10^3)$

$= 3 + 30$

$= 33$

4W SSPA $= 10^3$ mW/1mW

$= 10 \log 4 + 10 \log (10^3)$

$= 6 + 30 = 36$ dBm

10W SSPA $= 10 \log_{10} + 10 \log_{10} (10^3)$

$= 10 + 30 = 40$ dBm

25W SSPA $= 25 \times 10^3$ mW

$= 5 \times 5 \times 10^3$

$= 7 + 7 + 30 = 44$ dBm

Return loss: Return loss in dB can be expressed as per microwave circuit theory:

Return loss $= -20 \log |\rho|$

ρ is the voltage reflection coefficient and for

$\rho = 1$, Return loss $= 0$ dB

$\rho = 0$, Return loss $= \infty$ dB

Group delay: It is the delay of the complete signal when group of signals pass through a transmission system and defined as:

$$\tau(w) = -\frac{d\phi(w)}{dw} = -\frac{d}{dw} S_{21}(jw)$$

$\phi(w)$ – phase; S_{21} –transmission coefficient of two-port network

ω–angular frequency

It is also considered delay time for the signal envelope

$$\tau(w) = \frac{4Q_e}{w_e\left(1 - \frac{Q_e}{Q_0}\right)} \quad \text{for coupling structure}$$

$$\tau(w) = -\frac{d}{dw} avc\tan\frac{I_m(S_{21})}{R_e(S_{21})}$$

The slope of the phase frequency characteristics differentiate between positive group delay (PFD) or negative group delay

(NGD). Negative slope leads to PGD whereas positive slope leads to NGD ($Q_e > Q_0$).

Source: *IEEE Microwave Magazine*, Feb. 2021.

Annexure 2

S-parameter

S-parameters are essential due to the following reasons

- Non-availability of terminal voltage or current measurement (at high frequencies filed operations)
- Achieving short or open circuit in wide range of frequencies are difficult
- Instability of the circuit in short or open conditions in case of active circuits

General two-port model is associated with the set of four variables where two variables represent the excitation of the network (independent variables) and the remaining two represent the response of the network to the excitation (dependent variables). The linear network excited by terminal voltages V_1 and V_2 the terminal currents I_1 and I_2 are related by the following equations.

$$b_1 = S_{11}a_1 + S_{12}a_2 \qquad \text{... (A2.1)}$$

$$b_2 = S_{21}a_1 + S_{22}a_2$$

where the

$$a_1 = \frac{V_1 + I_1 Z_0}{2\sqrt{Z_0}} = \frac{V_{i1}}{\sqrt{Z_0}}$$

$$a_2 = \frac{V_3 + I_2 Z_0}{2\sqrt{Z_0}} = \frac{V_{i2}}{\sqrt{Z_0}} \qquad \text{... (A2.2)}$$

$$a_1 = \frac{V_1 - I_1 Z_0}{2\sqrt{Z_0}} = \frac{V_{i1}}{\sqrt{Z_0}}$$

$$a_2 = \frac{V_3 - I_2 Z_0}{2\sqrt{Z_0}} = \frac{V_{i2}}{\sqrt{Z_0}}$$

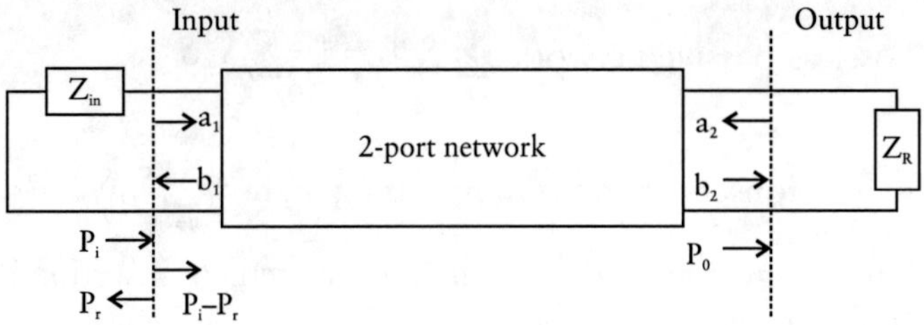

The S-parameters are

$$S_{11} = \left.\frac{b_1}{a_1}\right|_{a_2=0} = \frac{Z_i - Z_0}{Z_i + Z_0}$$

$$S_{22} = \left.\frac{b_2}{a_2}\right|_{a_1=0}$$

$$S_{21} = \left.\frac{b_2}{a_1}\right|_{a_2=0}$$

$$S_{12} = \left.\frac{b_1}{a_2}\right|_{a_1=0}$$

Two-port S-parameters are associated with the insertion loss/attenuation/gain (S_{21}) and return loss/voltage standing-wave ratio (VSWR) data (S_{11}).

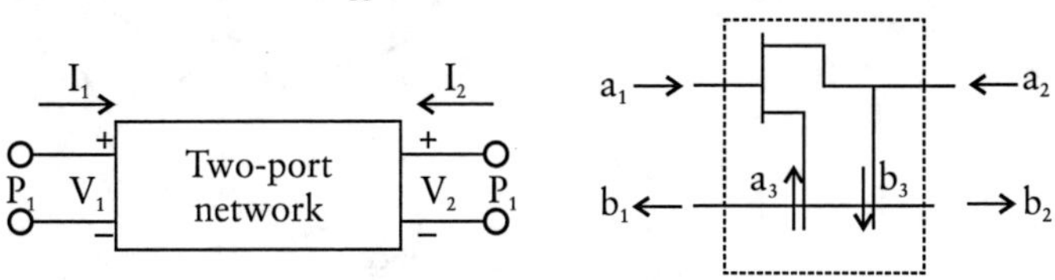

Two-port representation of FET

Important characteristics of S-matrices are:

S-matrix is symmetrical for reciprocal network

Lossless passive network $\sum_{n=1}^{N} |S_{ni}|^2 = 1 = \sum_{n=1}^{N} S_{ni} S^*_{ni}$

Power conservation yielding orthogonality $\sum_{n=1}^{N} S_{ni} S^*_{nj} = 0$

S-parameters: frequency domain measurement insertion loss, crosstalk, noise

Time domain: clock jitter, BER, interference, signal attenuation, inter-symbol interference (ISI)

S-matrix of come common circuits

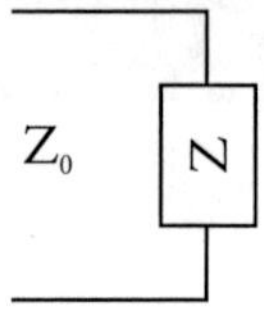

$$\frac{Z - Z_0}{Z + Z_0}$$

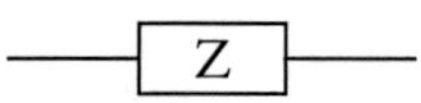

$$\frac{1}{Z + 2Z_0}\begin{bmatrix} Z & 2Z_0 \\ 2Z_0 & Z \end{bmatrix}$$

Smith Chart for transmission line representation

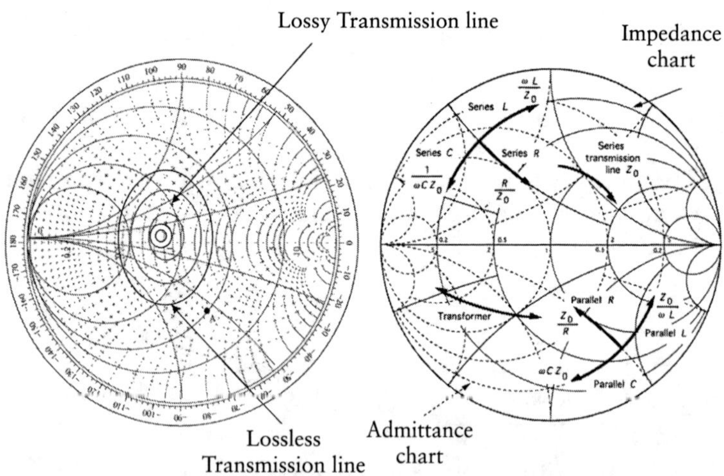

Fig. A2.Z-Y: Smith chart and impedance matching concept

Differential signal is used in high-speed digital applications. The parameters for differential S-parameters reveals differential return loss (SDD11), insertion loss (SDD21), near-end cross talk (SDD 31, SCC 31SCD31), far-end cross talk (SDD41, SCC41) and differential-to-common mode conversion (SCDxx).

Time-domain transmission (TDDxx) parameters are also vital for the characterization for jitter analysis (eye diagrams) and cross talk. It can be performed with an oscilloscope (TDR mode), bit error rate tester or VNA performing inverse discrete Fourier transform (DFT) to convert from the frequency domain (SDD21) to the time domain (TDD21).

Harmonic Balance Method: The periodic steady state of non-linear components is determined by global properties of the linear sub networks at discrete harmonic frequencies. The linear network is characterized at the harmonic frequencies. The analysis carried out using the following steps:

- Linear subnetwork reduced from the poles of non-linear components
- Determination of the steady state voltages and current at poles of the non-linear components of the networks
- Incident and reflected waves at the output ports of the network is determined by the steady state currents and voltages determined at poles of the non-linear components.

Sources: Januz Dobrowolski, *Computer-aided Analysis, Modelling and Design of Microwave Networks*, Artech House, 1996

Iulian Rosu, YO3DAC/VA3IUL

D.K. Misra, *RF and Microwave Communication Circuits*, Wiley Inter-science 2004

Annexure 3

Wave Propagation Mechanism in Microstrip

The major losses in microstrip can be categorized into:

(a) Conductor loss

(b) Dielectric Loss

(c) Radiation loss

Conductor loss is encountered at RF frequencies and is different from the DC losses as resistance of the metal increases with the frequency. As RF frequency increases, the current tends to flow nearer the surface, or the 'skin' of a conductor. The conductor loss is represented by the attenuation constant α_c which is given as

$$\alpha_c = \frac{8.686}{Z_0 w}\sqrt{\frac{\Pi f \mu}{\sigma}} dB/m \quad \text{... (A3.1)}$$

Dielectric loss also plays an important role in the overall performance of the circuit. The dielectric loss associated with the circuit depends on the permittivity and the loss tangent associated with it and represented as

$$\alpha_d = (27.3)\frac{\varepsilon_{eff} - 1}{\varepsilon_r - 1}\frac{\varepsilon_r}{\varepsilon_{eff}}\frac{\tan\delta}{\lambda_g} \quad \text{... (A3.2)}$$

Radiation loss is mostly associated in microstrip due to the inhomogeneous medium encountered due to quasi-TEM propagation. The main reason associated with the radiation is the insufficient time for magnetic field to build up causing loss of energy. This loss of energy radiates out from the conductor as an electromagnetic wave.

Conductor loss at low frequency remains constant and at higher frequencies it increases as skin effect concentrates current near the conductor surface and edge effect further restricts current to the edges. At high frequency, surface resistance can be shown as:

$$R_s = \frac{\rho}{\delta} = \sqrt{\pi f \rho \mu_0} \text{..........} t > 3\delta$$

$$= \frac{2\rho}{t} = \text{..........} t < 3\delta$$

$$R_{DC} = \frac{1}{\sigma t} = \frac{\rho}{t}$$

Transmission line with the thickness t >3 has the resistance of $\frac{Rs}{2}$ because both the surface of the transmission line contribute to the surface resistivity.

The frequency at which conductor transition from electrically thick to electrically thin frequency can be represented as:

$$f_c = \frac{4}{\pi\mu\sigma^2} \text{ in Hz}$$

The loss increase due to edge singularity and shown as f_c= R/(2ΠL). At the critical frequency f_c conductor thickness equals twice the skin depth.

Source: James C. Rautio, *Microwave Journal*, 2000

Static TEM parameters for Microstrip

$$\varepsilon_{eff} = \frac{C_d}{C_a} = \left(\frac{c}{v_p}\right)^2 \quad ; \; Z_0 = \frac{1}{c\sqrt{C_a C_d}}$$

$$\varepsilon_{eff} = \frac{1+\varepsilon_r}{2} + \frac{\varepsilon_r - 1}{2}\left(1 + 12\frac{h}{W}\right)^{-0.5} ; \lambda_g = \frac{\lambda_0}{\sqrt{\varepsilon_{eff}}} \qquad \text{... (A3.3)}$$

$$Z_0 = \frac{\eta}{2\pi\sqrt{\varepsilon_{eff}}} \ln\left[\frac{A}{x} + \sqrt{1 + \left(\frac{2}{x}\right)^2}\right]$$

$$A = 6 + (2\pi - 6)\exp\left[-\left(\frac{30.66}{x}\right)^{0.7528}\right]; x = W/h$$

For wide microstrip lines

$$\varepsilon_{\text{eff}} \to \varepsilon_r$$

For narrow lines

$$\varepsilon_{\text{eff}} \cong \varepsilon_r + 1/2$$

For coupled lines (even/odd modes)

$$Z_0 = \frac{\eta}{2\pi\sqrt{\varepsilon_{eff}}} \ln\left[\frac{A}{x} + \sqrt{1 + \left(\frac{2}{x}\right)^2}\right]$$

$$\varepsilon_{eff,e} = \frac{C_{ed}}{C_{ea}}; \varepsilon_{eff,o} = \frac{C_{od}}{C_{oa}}$$

Transverse resonance mode

$$f_c = \frac{c}{\sqrt{\varepsilon_r}(2W + 0.8h} \qquad \text{... (A3.4)}$$

where W is width of transmission line and h is height of the substrate.

Concept of free space

- Z_0 = 377 Ω (em wave propagation)
- No mechanical resistance
- Neither short or open circuit for e m waves
- Relative permittivity and permeability are 1

Skin Depth $= \sqrt{\frac{p}{\pi f \mu}} = \frac{1}{\sqrt{\pi f \mu}}$; where σ is the resistivity

Permittivity: It represents the easiness of material polarization due to displacement of charges

Permittivity: It represents the magnetic properties of the material and easiness of the magnetization of the material.

Impedance: The impedance for the transmission line

$$Z_0 = \sqrt{\frac{R + jwL}{G + jwC}}$$

In case of high frequencies or infinite line length the equation becomes real and can be represented as $Z_0 = \sqrt{\frac{L}{C}}$

The other representation are $Z_0 = \sqrt{\frac{E}{H}} = \sqrt{\frac{\mu}{\varepsilon}}$; in case of free space 377 Ω.

In radio wave propagation electro-magnetic waves suffer several effects resulting in power loss

Reflection and refraction: Polarization of the incident waves and 180° phase shift in the reflected component. This phenomena is responsible for large scale fading.

Diffraction: This causes excess path length resulting in phase shift and Fresnel zone relates phase shift to the position of obstruction. This phenomena is responsible for small scale fading.

Scattering: Incoming waves scattered into several weaker outgoing signals when the objects are smaller than the wavelength of the propagation waves (street signs, rain droplets, foliage, concrete walls, lamp posts). This phenomena is responsible for small-scale fading.

Link budget gives an account for all the gains and losses from transmitter through a medium to the receiver.

Table A3.1: Radio frequency bands

Frequency bands	Frequency Range	Wavelength range
VLF	3-30 kHz	10-100 km
LF	30-300 kHz	1-10 km
MF	300-3000 kHz	100-1000 m
HF	3-30 MHz	10-100 m
VHF	30-300 MHz	1-10 m
UHF	300-3000 MHz	10-100 cm
SHF	3-30 GHz	1-10 cm
EHF	30-300 GHz	1-10 mm
THF	300-3000GHz	0.1-1 mm

Noise power related with the bandwidth as per below relations which shows that higher bandwidth result in the higher noise power.

Noise power = power spectral density (dBm/Hz) × 10 log (BW)

Table A3.2: Microwave frequency bands and applications

Frequency bands	Applications
850MHz-1.9GHz	Wireless
4GHz-6GHz	Satellites
12GHz	PAA, DBS
20GHz	Satellite downlinks
27GHz-35GHz	Missile seekers, 5G, LMDS
44GHz (Q-band)	SATCOM terminals
60GHz (V-band)	Wireless LAN, MVDS
77GHz	Collision avoidance radar
94GHz (Wband)	FMCW radar

Source: B. Preetham Kumar, *Communication System Laboratory*, CRC Press, 2016

Annexure 4

Cables and Connectors

Cables and connectors are the basic elements in radio frequency employed for the measurement and characterization of various circuits and components, device under test (DUT). They provide electrical contact between testing equipment and device under test (DUT) and depends on the signal characteristics, frequency range and band width.

Parameters effecting RF measurements are:

- Connectors mating/lossy connector
- Uncertainty (instrumentation, calibration, mismatch, loss compensation)
- System noise
- External/unwanted signals
- Temperature
- Impedance
- Device gain and noise figure
- Jitter display
- Adaptors/connectors precautions (tightening/cleaning)

A. Connectors

The size of the connectors shrinks as the frequency increases. The earlier invented UHF plugs and jacks are usable uptil 30 MHz which are replaced by BNC which are comparatively fast and convenient. BNC (Bayonet Navy connector) connectors are of various types such as slide on, polarized, twinax, triax etc. and can be used from DC-40GHz conveniently Also they can handle power levels from pW-kW levels. N-type and 7 mm connector enhances bandwidth upto 18 GHz and exhibits low insertion loss, high reliability apart from wide bandwidth. Subsequently, SMA and APC (Amphenol precision connector)-3.5 connectors are extended the frequency range upto 30 GHz. The important aspect in connectors is the frequency range and are shown as:

Table A4.1: Connectors for various frequency ranges

Connector type	Frequency Range
UHF	300 MHz
BNC	4 GHz
3.5 mm	18 GHz
2.4 mm	50 GHz
1.85 mm	65 GHz
1.0 mm	>110 GHz

The main aspect in the connector is the characteristics impedance Z_0 which can be represented as $Z_0 = \sqrt{\frac{L}{C}}$.

Any physical discontinuity in the connector can affect the L and C which are defined per unit length. The usage of connectors beyond recommended frequency range will lead to multi-moding or higher VSWR.

Connector Construction: The connector material is passivated with stainless steel and BeCu plated with Ni/Cu. The above assembly is gold to have high surface quality, conductivity and better corrosion resistance.

Table A4.2: Connectors material

Material	Usage	Applications
Passivated stainless steel	Coupling nuts & outer body	• Corrosion resistance • Improved mating-de-mating cycles
BeCu	Conductor pin	• Dimensional stability over temperature
Ni	Base/Adhesive layer	• Enabling gold plating
Au	Outer plating	• Good conductivity • Corrosion resistance

Further for enhanced stability, the technique of stain relief is incorporated. General Connector Overview:

1. 7 mm connector: coupling nut moved clockwise, extends coupling sleeve and it becomes female.

Counter clockwise movement, it retracts and become male.

2. Female connectors types: receptable, slot, outlet, socket

Male connectors types: plug, pin, prong

3. Calibration Techniques-SOLT, TRL/LRL, TRM/LRM

4. Push pull connector-cable interconnects that provide a string locking mechanism which is released by squeezing the connector body.

5. Blind mate connector-mating action without seeing or feeling.

Coaxial Connectors

Coaxial connector are employed for interconnections and presently preferred over waveguide at mm-wave frequencies. Waveguide was employed for interconnections but lacks broadband frequency coverage and single sweep measurement while coaxial connectors possess single sweep, wide bandwidth, frequency scalability and avoidance of uncertainty due to impedance variations.

Table A4.3: Connectors for various frequency ranges

Electrical Parameters	Mechanical Parameters
Frequency coverage	Design of connector meeting repeatability and mating, Dimensions
Impedance	Nut tolerances

RF connectors

$$Z_0 = \frac{1}{2\pi}\sqrt{\frac{\mu}{\in}}\ln\left(\frac{D}{d}\right)$$

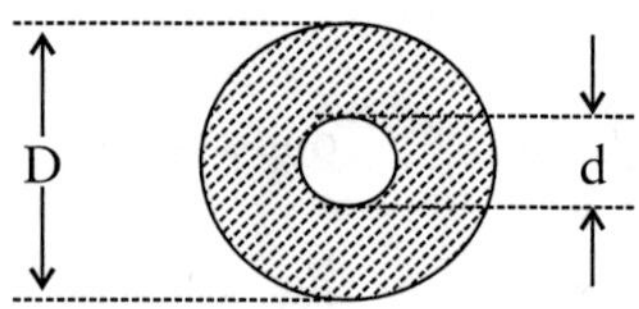

D = inner diameter of outer card

d = outer diameter of inner card

Higher order modes occur $\left(\frac{D+d}{2}\right) \geq \lambda$

Using the formulae:

$$f_c = \frac{c}{\lambda_c\sqrt{\mu_r \varepsilon_r}}$$; c-speed of light, ε_r-relative permittivity

$$Z_0 = \frac{1}{2\pi}\sqrt{\frac{\mu_r}{\varepsilon_r}}\ln\frac{D}{d} = 130\log\frac{D}{d};$$ where D = inner diameter of outer conductor, d=outer diameter of inner conductor

Table A4.4: RF connectors overview

RF connectors (mm)	Cut off freq (GHz)	Impedance (Ω)	Center Conductor (mm)
3.5	40	33	1.52
2.92	46	40	1.27
2.40	56	50	1.04
1.85	73	70	0.80
1.0	133	110	0.43

B. Cable

Coaxial lines are available as flexible, semi-rigid and rigid cable and can be employed upto 100 GHz. Flexible coaxial line consist of center conductor and an outer shielded connector which is protected by an outer cover. In semirigid outer conductor is having solid tubing which is used for connecting short lengths along with the complete shielding protection. Rigid coaxial cable has solid outer conductor having air as dielectric along with spacers/beads to separate conductor from outer conductor. The attenuation for RG 214 cable at 1 GHz is 0.3 dB/m which increases to 1.57 dB/m at 10 GHz.

The size of cables at higher frequencies shrinks in size and the reduction in the diameter of coaxial cables yields higher insertion loss per unit length at higher operating frequencies or wider bandwidth. This phenomena can be compensated with the selection of appropriate connectors.

Main specifications of the cable are:

Impedance, frequency range, dielectric breakdown voltage, insulation resistivity, phase stability(with bending), amplitude stability, VSWR, velocity of propagation, capacitance, shielding effectiveness

The commonly employed cables are RG178, RG 214, RG 217 having 50 impedances and outer dimensions are 0.071″, 0.425″, 0.545″. The important considerations in implementing connectors at RF frequencies are:

– cleanliness of connectors (removal of dirt, oil, etc.)

– tightening using torque wrench

The main losses associated with the cables are associated with the transmission of multi mode or higher order modes which can enhance attenuation and reduce power handling capabilities. The spacing between conductors effect the mode of transmission in the cables.

Requirement of coaxial cables at high frequencies for test and measurement are for:

- Ruggedness
- Reliability
- Consistency
- High performance

The main construction of the coaxial cable is the centre conductor dielectric and outer conductor. The return current travels through the outer conductor. The ruggedness and insertion loss are contradictory as ruggedness enhances the losses.

In case of high end applications such as aerospace and military which have subjected to vibration, shock, repeated flexure mating-unmating cycles etc requires cable to maintain consistent performance and calibration. The bottlenecks associated with the cable quality are:

Ripples in reflection and transmission magnitude

- Magnitude and phase errors
- Non-repeatable measurements
- Unstable measurements

The signal attenuation in the cable assembly is caused due to conductor and dielectric losses which are measured as insertion loss. The frequency dependent phenomena associated with the cable are: conductivity, skin depth, mode of propagation, propagation velocity, wavelength. The dimensions of cables become smaller at higher frequencies resulting in higher resistive losses and heat built-up under load. Beside frequency, humidity, temperature and other environmental factors also impact the cable performance.

The construction of the cable and associated connector selection are important considerations in achieving repeatable performances. This is due to low reflections and associated VSWR changes due to various phenomena such as vibration, hock, stress, repeated reposition flexure etc. The susceptibility of cable on the handling beside connector performance which can deviate due to oils, wear, contaminants affects their measurement repeatability.

The performance of cable assembly depends on the parameter-connectors, connector termination, and the cable. Coaxial cable assembly is characterized by time domain reflectometry and VSWR measurements.

The parameter of cable assembly depends upon the three parameters-connectors, connector termination and cable. The another aspect in the cable assembly is the shielding effectiveness which is dictated by the outer conductor. The shielding mechanism is based on two methodologies-reflection and absorption. At higher frequencies, the shielding mechanism is absorption which depends on material conductivity and permeability.

Table A4.5: Cable types and shielding effectiveness

Cable type	Frequency range (GHz)	Shielding effectiveness @ 18 GHz (dB)
Braided round-wire	1-18	40
Braided flat-wire	1-18	85
Helically wrapped flat wire	1-18	120
Helically wrapped metallized polymer foil	1-18	30

The semi-rigid cable due to continuous single piece construction provides shielding of > 140 dB from 1-18 GHz. The main selection parameters for cable assemblies are:

- Repeatable connections
- Rugged jacket
- Shielding
- Return loss
- Junction reinforcement
- Insertion loss and phase stability

The phase delay in coaxial cable can be given as:

$$\phi(f) = -360°f\tau \text{ (phase linear function of f)} \quad \text{... (A4.1)}$$

where f is frequency; τ is the time delay

$$\tau = \frac{L_{electrical}}{c} = \frac{L_{mech}\sqrt{\varepsilon}}{c} \quad \text{... (A4.2)}$$

where L_{elec} is the electrical length and L_{mech} is the mechanical length

The phase change depends on the electrical length, temperature variations, frequency and mechanical stress (flexing, vibration). So phase stable cable are constructed employing specialized dielectrics such as polytetrafluoroethylene (PTFE) having lower r, high flexure and decreases with the increasing temperature. A unit section of coaxial cable having resistance, capacitance and inductance which can be represented as

$$L = 0.0046\log_{10}\frac{b}{a}$$

$$R = 4.1\times10^{-8}\sqrt{f}\left(\frac{1}{b}+\frac{1}{a}\right)$$

$$C = \frac{0.24\varepsilon}{\log_{10}\left(\frac{b}{a}\right)}$$

$$Z_0 = \frac{138}{\sqrt{\varepsilon}}\log_{10}\frac{b}{a}$$

$$E = \frac{V}{r\ln\frac{b}{a}}$$

The maximum power carried by coaxial line is achieved by decreasing Z_0 whereas for maximum voltage the value of Z_0 to be increased. So the optimum value of 50 is generally kept to compromise the requirement of high power and high voltage.

Cable construction consists of five parts namely center conductor, dielectric, outer conductor, outer shield, and outer jacket. The outer conductor material nowadays employed is ARACON (polyparaphenylene terephthalamide) and outer jacket is either fluorinated ethylene propylene (FEP) or ethylene tetrafluroethylene (ETFE).

Sources:

1. Cables & Connectors, *Microwave Journal*-2019.
2. *Introduction to Microwaves*, Wheeler.

Annexure 5

Concept of Linearity and Breakdown

Linearity is an important consideration for preserving pulse in radar applications. It also effects the characteristics in radar and quality of modulation in communication system mainly in transmitter due to linearity of power amplifier, frequency mixers, switches, etc. Non-linearity results in the generation of IMD (inter-modulation distortion).

Secondary Distortion: Increase in IMD by 2-dB for every 1-dB increase in the input power whereas in third order distortion, increase of 3-dB with 1-dB increase of input power. IP2 and IP3 are only the mathematical representation.

Non-linearity results in signal degradation as well as degradation of modulation quality. Efficiency is highest at transmitter saturation so there is trade off between linearity and efficiency. Mixer linearity is characterized in terms of three order intercept.

Various linearity techniques are:

- Feed-forward
- Envelope tracking
- Digital pre-distortion
- Analog pre-distortion

RF Breakdown

The production of electrons in the intermediary gas exceeds removal of electrons. It can be categorized as:

1. Ionization breakdown: Gas ionization due to electron collision

This phenomenon happens when electron free mean path becomes smaller than the separation distance which is caused at higher ambient pressure.

2. Multipaction: Secondary electron mission

The electron mean free path is longer than the electron separation distance and occurs at ambient pressure of $\sim 10^{-7}$

Corona Effect

The close air gap between electrodes/conductors leads to high electric field sensitive to ESD. Applied voltage within order of magnitude of characteristics voltage, the real part of the E within the gap can differ many order of magnitude from the predicted E without considering the static charges.

Annexure 6

Concept of Quality Factor

The concept of quality factor is an important as it effects the bandwidth and insertion loss of the circuit. This concept is important in oscillator applications.

Basically for generalized transmission lines, quality factor is an important parameter, given by

$Q_u = \pi/\lambda_g \alpha$; where α is attenuation constant

Above relation demonstrate that variation in unloaded quality factor depends in loss (α) as well as on dielectric constant (λ_g).

$$\frac{1}{Q_l} = \frac{1}{Q_u} + \frac{1}{Q_e}$$

$$\frac{1}{Q_e} = \frac{1}{Q_R} + \frac{1}{Q_c}$$

$$Q_u = \frac{1}{\tan \delta}$$

$$Q = \frac{Q_e}{1 + Q_E \tan\delta}$$

Q_u = unloaded quality factor and limited by the conductor and dielectric losses

Q_R = Q-factor related with reactive coupling between resonator and feeding CPW regions

Low resistivity substrate such as silicon having high substrate loss and the unloaded quality factor can be represented as

$$Q_u = \frac{f_0}{\Delta f \times \left(1 - |S_{21}|\right)}$$

$$Q_u = \frac{Q_L}{\left(1 - 10^{-\frac{IL}{20}}\right)}$$

In case magnitude of S_{21} is very small and approaches to zero than

$$Q_u = Q_L = \frac{f_0}{\Delta f}\text{(weakcoupling)}$$

$$Q_r = \frac{\omega W_T}{P}$$

$$W_T = \frac{1}{4} h A \varepsilon_0 \varepsilon_r$$

$$Q_C = h\sqrt{f}\pi\mu\sigma$$

$$Q_d = \frac{1}{\tan\delta}$$

$$\frac{1}{Q_T} = \frac{1}{Q_r} + \frac{1}{Q_d}\frac{1}{Q_{sur}} + \frac{1}{Q_r}$$

$$\frac{1}{Q_T} = \frac{P_d + \mathrm{P_r} + P_{sur} + P_c}{\omega W_T}$$

Resonator Q-factor measurement can be carried out using the formula

$$Q = \frac{S_{21}(f_0)}{S_{21}(3dB)} = \frac{f_0}{f_1 - f_2}$$

The peak of S_{21} or S_{11} is taken as resonant frequency. In either case, the other parameter can be found out using the S-parameter theory for a lossless circuit

$$S_{11}^{\ 2} + S_{21}^{\ 2} = 1$$

In case of weak coupling having poor return loss, the Q-factor could be deduced by the S_{21} using the above formulas.

Source: B. Walker, "Determining Resonator Q-factor from Return Loss measurement alone," *Microwaves & RF*, July 2020.

Annexure 7

Microwaves for Communication

Microwave communication are used for short range communication which are ideal for TV and audio while satellite communication can be established for long distances such as ships, aircrafts, relaying telephone calls.

Microwave can be focussed and received via aerial antenna and can extend range further beyond visible horizon. Receiver is placed on top of high buildings or hilltops and mountain peaks because higher the receiver is farther broadcast of signal can be broadcasted.

Satellite communication can reach all areas of earth. It provide satellite communication links to direct broadcast, satellite phone service and individual satellite communication links.

Table A7.1: Frequency bands for communication systems

Applications	Frequency
AM Radio	LF: 153-279 kHz
	MF: 531-1611 kHz
	HF: 2.3-26.1 MHz
FM Radio	76-108 MHz
Satellite	1-75 GHz

Line of Sight (LOS)	1-25 GHz
Cellular	800 MHz-2.5 GHz

Receiver Sensitivity: The main receiver sensitivity can be calculated as

Required sensitivity (min.)=P_r + SNR

Receiver noise input power $P_r = F + 10 \log (kT_0B)$

where F is the receiver noise figure (dB), k is the Boltzmann constant (1.38×10^{-23} W.S/k), T_0 is the absolute temperature and B is the receiver noise bandwidth (Hz)

Receiver sensitivity and required SNR depends on the modulation type and coding rate. Link budget in mobile communication takes shadowing and multipath effects. Microwave communication is LOS as they get attenuated when used as ground waves as well as surface waves.

Link margin is an important concept for the reliable satellite link. In digital communication, required signal level is defined as a ratio of energy per bit to receiver noise level (E_b/N_0). The required signal level is a function of BER for the signalling scheme and represents the probability of a bit error rate at the output of the receiver. Received power at receiver translated to energy per bit as:

$[E_b/N_0]_r = P_r \times [W/R]$

Link Margin=$[E_b/N_0]_{received}-[E_b/N_0]_{required}$

Microwave communication are used for short range communication which are ideal for TV and audio while satellite communication can be established for long distances such as ships, aircraft, relaying telephone calls.

Microwave can be focussed and received via aerial antenna. Receiver is placed on top of high buildings or hilltops and mountain peaks because higher the receiver is farther broadcast of signal can be broadcasted.

Table A7.2: Frequency bands for communication systems

Applications	Frequency
AM Radio	LF: 153-279 kHz
	MF: 531-1611 kHz
	HF: 2.3-26.1 MHz
FM Radio	76-108 MHz
Satellite	1-75 GHz
Line of Sight (LOS)	1-25 GHz
Cellular	800 MHz-2.5 GHz
Satellite Communication	S-, C-, X-, Ku-, Ka-band

Annexure 8

High Frequency Printed Circuit Board

Hybrid integration is classified into three types based on pattern methodology and are classified as: printed circuit board,thick film substrate and thin film substrate.

The circuit operating from 100 MHz-2GHz are considered to be RF printed circuit board whereas circuits above 2 GHz are considered as microwave PCB. PCB for RF applications deals with the transport of energy using transmission line which can be microstrip, suspended stripline, coplanar waveguide.

RF PCB is designed either employing 2 or 4 layer board where top layer is employed for components and signal routing whereas bottom layer for grounding. The continuity and shortest ground path route to be ensured for the return current path. Long ground return path can cause parasitic inductance leading to unintended feed back loop.

The ground planes connected using vias to be added continuously throughout the RF portion preventing ground inductance due to ground return paths which also facilitate in cross coupling.

RF boards contain broadly outer laminated layer and the core of the board. The outer laminated layers (prepreg) of PCB contain less glass content compared to the core. Mostly FR4 core is employed having $\varepsilon_r = 4.2$ whereas outer laminate having

$\varepsilon_r = 3.8$. The metal is generally copper having thickness of 1 oz (1.4 mils).

Important considerations in RF boards are:

- Board selection
- Line impedance (mostly 50 ohms)
- Line width
- Substrate height
- Mitering smoothness
- Routing of line as far as possible
- Power lines to be separated from the signal lines
- Common ground path

Trace impedance at RF frequency depends on:

- Dielectric Constant
- Board Thickness
- Board Height

RF-ASICs such as prescalar VCO needs solid ground plane beneath the component placement. The ground paddle carries RF and DC return currents and act as thermal heat sink. The heat sink property can be enhanced by filling the vias with conductive paste.

Advantages and disadvantages of multilayer construction for digital PCBs

(a) Advantages

1. Low-impedance power distribution.
2. More uniform Z_0 for signal tracks.
3. Reduced EMI susceptibility because of low circuit loop area (all conductors are close to the ground plane).
4. Reduced EMI emission because of slower edges on logic transitions, caused by distributed capacitance, and low loop area.
5. Reduced crosstalk among tracks because of proximity to ground plane.

(b) Disadvantages

6. Higher manufacturing costs compared to double-sided plated-through hole (PTH) boards.
7. Difficult to rework and repair.

Advantages and disadvantages of double-sided construction for digital PCBs

(a) Advantages

1. Low manufacturing cost.
2. All connections accessible for rework and repair.

(b) Disadvantages

3. Careful attention to layout needed, especially in power distribution.
4. Susceptible to radiated EMI because of open structure of circuit loops.
5. Significant EMI emissions from fast edges and clocks.
6. Crosstalk between adjacent closely spaced tracks.

In high frequency board the choice of the PCB material is based on the factors such as cost, performance, applications and at RF the primarily parameters are dielectric constant, dissipation factors, thermal coefficient of dielectric constant, thermal conductivity etc. Commercially high frequency PCB material can be categorized in seven types as per Table A8.1.

Table A8.1: PCB substrates

Substrate	Cost	Metallization	Machinability
Ceramic	Medium	Fair	Poor
Plastic	Low	Poor	Fair
Silicon	High	Good	Very good
Glass	Low	Good	Poor

In high frequency board the choice of the PCB material is based on the factors such as cost, performance, applications and at RF the primarily parameters are dielectric constant, dissipation factors, thermal coefficient of dielectric constant, thermal conductivity etc. Commercially high frequency PCB material can be categorized in seven types as per table.

Table A8.2: High frequency PCB material

Material	Circuit Fabrication	Electrical Performance
PTFE with micro glass fibre	Difficult	Excellent
PTFE with woven glass	Difficult	Good
Ceramic filled PTFE	Moderate	Excellent
Ceramic filled PTFE with woven glass	Moderate	Good
Ceramic filled Hydrcarbon	Difficult	Good
Ceramic filled Hydrocarbon with woven glass	Easy	Good
High performance FR-4	Easy	Poor

The main concern during PCB assembly relates with the thermal stress from soldering, solder rework or multiple thermal cycles. The CTE value affects the performance and circuit material with lower CTE is more robust than the higher CTE material. This leads to the choice of ceramic filled PTFE laminate with woven glass having excellent electrical performance and good CTE. Micro-fibre PCB provides better electrical performance whereas ceramic filler improved the thermal conductivity. Overall low dissipation factor with robustness achieved for the PCB.

Table A8.3: RF Boards and properties

RF Boards	Permittivity	Properties	Assembly
Glass Epoxy	4-4.25	• Less warpage • Solder and Gold plated variant	SMD & Thru hole components
TFG (soft substrate)	2.5	• Moderate warpage • Solder and Gold plated variant	SMD & Thru hole components
RT Duroid (soft substrate)	10.2	• More warpage • Gold plated variant	SMD, carrier plate or box mounted
Alumina	9.8	• No warpage • Gold plated variant	SMD, carrier plate mounted

The operating conditions and the end-use environment for which the circuit is subjected, can play a role in the variations of phase response and Design permittivity. Variations of phase response and permittivity are typically assumed to be mostly related to circuit material properties and their variations.

Multilayer PCBs manufacturing poses many challenges such as drilling, PTH preparation, CTE mismatch as materials differ in dielectric constant, dissipation factor and processing temperatures. Drilling process introduce many defects such as smear, burring, material fracturing.

Moisture absorption presents challenge of enhancement of dielectric constant and loss which impact the electrical performance.

The wire employed for connections can be categorized in AWG (American Wire Gauge) and the same can be represented as:

Table A8.4: Wire gauges and equivalent diameter

American wire gauge	Diameter (mils)
AWG50	1
AWG44	2
AWG38	4
AWG32	8
AWG26	16
AWG20	32
AWG14	64

Presently RF-CMOS is gaining popularity which is the extension of MMIC on silicon substrate.

Table A8.5: Silicon Vs GaAs

Parameters	Silicon	GaAs
Cost effectiveness	Low cost	High cost
Die size	Large	Comparatively small
Process	Well controlled micromachining	Complex process
Integration	Higher range	Incorporation of off-chip components

The active devices can be Si-Ge. The advantages associated with the Silicon-Germanium Devices are

Lattice mismatch between Si and Ge, so only light Ge doping

Si based so cheaper than GaAs

Lower performance compared to GaAs

Low breakdown voltages (enhanced instability in power amplifier)

Applications in high speed digital, wireless hand portables

In case of Silicon losses are higher due to finite resistivity so induced current in bulk silicon substrate is produced due to electro-magnetic coupling .Further em signal introduces alternative polarization of the dielectric molecules and collision of the crystal lattice.

The losses associated with the resistivity in case of CPW lines is tabulated in A8.6.

Table A8.6: CPW line on Silicon

Resistivity (Ω-cm)	Loss (dB/mm)
0.5	2.9
20	1.7
20 with additional layer of SiO (1 um)	0.7
20 (SOI wafer)	1.1

Role of Feedthrough

Feedthrough is considered as a filter employed in the circuits to prevent EMI from interfering with the devices. It is a component having capacitive element built around the central conductor mostly employing ceramic. So due to capacitance effect, RF energy at high frequencies are passed due to short circuit effect whereas low frequencies RF energy are not allowed due to open circuit effect.

The main requirements for the Feedthrough are

- Insertion loss
- Power

- Cut off frequency

Various types of Feedthrough filters are

Individual Line

- SMT Solder-in
- Press-in
- Bolt-in

Multiple Line

- D-sub
- Mil-circular
- Filter plate

Sources:

1. *Microwave Journal*, 2005
2. *Microwaves & RF*, May 2020

Annexure 9

Numerical Techniques

Filed solvers are employed to take into account parasitic coupling, interaction with the boundary nd discontinuity between the elements. Full electromagnetic analysis covers the aspect of design and synthesis. It involves to find the filed/source function that obey Maxwell equations, satisfy boundary conditions at all interfaces/material and satisfy all excitation functions. The methodology can be segregated into three main methodology namely: Analytical method (closed form expressions), Semi-analytical method (infinite series, variation formula), Numerical method (integral or differential form of Maxwell equations).

The numerical techniques can be broadly classified as

1. Integral equation method
2. Differential equation method
3. Ray tracing method

The technique selection is based on the geometry of the problem and the frequency range. The various techniques can be segregated as:

- Finite Difference Technique
- Variational Technique

- Finite Element Technique
- Method of Moments
- Mode Matching Technique

Method of moment technique is applied for the solution related with the circuit whose maximum physical dimensions are the order of wavelengths over the frequency of interest. This technique can be applied to integral equation formulation as well as to differential equation formulations. Maxwell's electromagnetic equations are transformed into integral equations with suitable electric and magnetic Green's function applied in the multilayered substrate. This formulation expresses the electric and magnetic field as a combination of a vector and a scalar potential. The unknowns are the electric and magnetic surface currents flowing in the planar circuit. The method of moments procedure reduces an operator equation of the form

$$Lf = g$$

L-linear operation

g-known function

f-unknown function and can be represented as $f = \Sigma\alpha_j f_j$

where f_j are known as basic functions and j are scalars to be determined. So the equation can be represented as:

$$\Sigma\alpha_j f_j L = g$$

A set of testing or weighing functions in the range of L can be defined as

$$\Sigma\alpha_j \langle w_i f_j L \rangle = \langle w_i, g \rangle; \text{ where } i = 1, 2, 3, 4, \ldots\ldots$$

The main step in the method of moments technique is the choice of basis functions which can represent the unknown functions with accuracy and minimizing the computational effort. The unknown solution is expanded in terms of known solution with unknown coefficients and can be solved using 1D, 2D, 2.5D, 3D method accordingly.

In case of regular structures, whole domain functions are used and common entire-domain basis set is sinusoidal functions. Sub-domain basis functions are employed for the complex structures and commonly employed functions are pulse functions, triangular functions, piecewise sinusoidal functions. In the case of arbitrary shaped structure, triangular patch modelling approach is adopted having specialized basis function. A set of weighting functions is defined over the discretized domain and an inner product of each testing function with the boundary condition equation is formed.

The sources applied at the ports of the circuit yield the excitations in the equivalent network model. The currents in the equivalent network are the unknown amplitudes of the rooftop expansion functions. Solving the equivalent network for a number of independent excitation states yields the unknown current amplitudes. The current solution yields the S-parameter data of the circuit.

So method of moment technique to an integral equation can be applied by adopting following steps:

1. Represent the physical structure by equivalent models in which equivalent sources are radiating into regions for which Green's functions are known.
2. Operator equation to be obtained for the equivalent sources and expanding the sources in terms of expansion (basis) functions with unknown coefficients.
3. Discretize the domain over which the equivalent sources exist and expand the sources in terms of expansion functions with unknown coefficients.
4. Enforce the boundary condition equation (Step 2) over the discretized domain to obtain algebraic equations.
5. Solve the system of equations obtained (Step 4) and obtain the unknown coefficients on the basis of set defined in Step 3.
6. Compute other quantities from the equivalent source distribution.

Mode Matching Technique

MMT include higher order modes excitations and interactions resulting in the contributions of evanescent TE and TM modes to the overall electromagnetic fields. The following steps for the calculation are:

1. Field is defined for the both sides of discontinuity in terms of incident and reflected signals of unknown amplitudes.

2. Field component are described in terms of fundamental and higher order modes.
3. Magnitude of average power carried by each mode taken to be unity.
4. Application of boundary conditions at the discontinuity.
5. Principle of mode orthogonality to isolate amplitudes of forward and backward travelling wave.
6. Deduce scattering matrix by setting algebraic equations and solving it.

Finite Difference Methods

Various FEM methodologies are: Galerkin, Least square, Spectral. The unknown quantity is field or potential. Integral equation can be represented as:

$$\int_a^b f(x)dx = \sum_{m=1}^{n} a_m f(x_m)$$

Subsequently the integral equation can be written as

$$f(t_i) = \sum a_m K(t_i, t_m) f(t_m) = \phi(t_i),\ i = 1, 2, \ldots n$$

t_1, t_2 ...t_n are points in which interval (a, b) is sub-divided. The above equation is a system of n linear equations with n unknowns.

FDTD is also technique which is finite difference over small sub-domain based on time domain. Maxwell equations are written in curl formula which are converted to central difference equations and solved in the time domain.

$$V(t, x_i) = \int_{C_v} \overline{E(t, x_i)}\,\overline{dl}$$

$$I(t, x_i) = \oint_{C_I} \overline{H(t, x_i)}\,\overline{dl}$$

V is the common contour extending from ground plane (voltage reference point) to the circuit location x_i which is pointing on a microstrip. The $V(t,x_i)$ is independent of the choice of C_v considering path confinement in transverse plane whereas contour C_I wrap completely around the strip conductor at its surface in the transverse plane providing local current. E and H are initialized at a given time after problem is taken into gridded form, than. Time is advanced one step and fields are computed again resulted in convergence when filed quantities reached to a steady state.

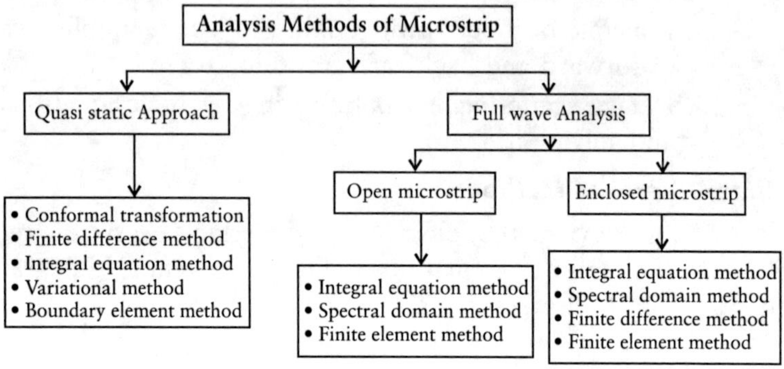

Fig. A9.1: Microstrip analysis techniques

2-D solvers are MOM, SDM whereas 3-D solvers are based on FEM, FDTD.

Source: Allen Taflere, S.C. Hagress, *Computational electro-dynamics: The finite difference time-domain method*, Artech House, 2000

Annexure 10

Frequency generation methodology using PLL Method

Typical methods

1. Direct synthesis
2. Indirect synthesis

Direct synthesis method:

- Stable reference source (TCXO, OCXO)
- Multiplier chain for desired frequency
- Better phase

Indirect synthesis

- Stable reference source
- VCO at desired frequency
- VCO frequency divided for comparison
- Poor phase

Differential Signals

Single ended signal (unbalanced):

- It is difference between signal of desired signal and a constant reference point
- Ground reference unaffected by the injected error which is carried forward through the signal
- Needs complex cancellation techniques to remove error

Differential Signal

- Pairs of balanced signals
- Signals moved at equal but opposite amplitude
- Error introduced in differential system path got added to each of the balanced signal equally
- Error cancellation resulting in better CMRR (common mode rejection ratio) and PSRR (power supply rejection ratio)
- Composite signal swing can be twice that of single-ended swing resulting in the enhanced SNR
- Same signal amplitude with lower supply voltage (less power dissipation) or amplification can b increased resulting in lowering of distortion
- Inherent cancellation of even-order harmonics in differential systems

Annexure 11

Material selection in MIC

Gold is preferable because it has very low contact resistance, high melting temperature and resists surface layer formation. Also it has advantage in wet sacrificial etching and can be electroplated easily.

Table A11.1: Comparison of Aluminum and Gold

Mechanical Properties	Aluminum (Al)	Gold(Au)
Young's Modulus (GPa)	70	78
Poission's Ratio	0.35	0.44
Residual Stress(MPa)	–20 to 160	–50 to –60
Density g/cm^3	2.7	19.3

Shear Modulus (GPa)	26	27
Yield Strength (MPa)	55-310	205
Knoop Hardness (kg/mm^2)	130	150
Thermal and Electrical Properties		
Coefficient of Thermal Expansion (ppm/K)	23.1	14.2
Thermal Conductivity @ 300 K(W/cm.K)	2.35	3.2
Heat Capacity (J/gK)	0.904	0.129
Melting point (^{0}C)	660	1064
Resistivity (Ω-cm)	2.65×10^{-6}	2.2×10^{-6}

Role of thermal conductivity on RF power handling capability can be explained by using below formulas:

$$\Delta T = \frac{0.2303h}{K}\left[\frac{\alpha_c}{w_{eff}} + \frac{\alpha_d}{2w_{eff}(f)}\right] {}^0\text{C/W} \qquad \text{... (A11.1)}$$

where w is the effective width, α_c and α_d are the conductor and dielectric loss.

$P_{mavg} = \theta/(\Delta T)$; θ is the temperature rise in ^{0}C ... (A11.2)

Lower value of ΔT is required for high power handling.

Table A11.2: MIC substrate overview

Material	Permittivity	Loss tangent	Thermal Conductivity (W/cm-K)
Hard substrate			
Alumina	9-11	0.0001	0.37
Sapphire	9.4,11.6	0.0001	0.46
Soft Substrate/PCB board			
FR-4	4.3-4.8	0.01-0.02	
RT Duroid-5880	2.2	0.0004-0.0009	

Table A11.3: Properties of the common semiconductor materials

Material	Permittivity	Electric field (V/cm)	τ(sec)	Energy gap (eV)	Structure	Resistivity (Ω-cm)
Si	11.8	3×10^{5}	2.5×10^{-3}	1.1.07	Zincblende	1000
GaAs	12.5	4×10^{5}	$\sim10^{-8}$	1.35	Zincblende	10^{8}
SiC	10.7	3.8×10^{6}	$(10\text{-}100) \times 10^{-9}$	2.403	Wurtzite	$>10^{12}$
GaN	9.5	$>10\times10^{6}$	$(10\text{-}100) \times 10^{-9}$	3.1	Zincblende	$>10^{13}$
InP	12.4	5×10^{5}		1.29	Zincblende	$>10^{7}$

Annexure 12

Various Interfaces for Measurement

I²C(inter IC): The I²C interface was introduced by Philips semiconductors in the early 1980's. The name I²C translates into "inter IC" since the idea was to allow easy communication between components residing on the same circuit board. Currently I²C is not only used within a single board, but also used to connect separate devices using a cable. Each device connected to the bus is software addressable by a unique address. I²C is a multimaster bus, meaning that multiple masters can initiate data transfer over the shared bus. The main advantage of the I²C interface is its simplicity. Only two bidirectional lines (serial data and serial clock) are required for communication. Disadvantages include relativity low communication speeds and the lack of automatic bus configuration.

RS-232: Rs-232 is another serial interface that is widely used in computer serial ports. The standard was introduced by the Electrical Industries Association and evolved from serving electromechanical teletypewriters to modern electronic devices and personal computers. While the standard recommends 25 pin connector, a three wire arrangement is often used when the full capabilities of RS-232 are not required. In the later case, communication is established via the transmit data, receive data, and ground pins. RS-232 can be a good choice if the synthesizer needs to be controlled from a personal computer but its main disadvantage is relativity low- speed so it is being replaced by the much faster USB and Ethernet connections.

USB (Universal serial bus): Universal serial bus (USB) is today's most popular way of connecting various devices to a personal computer. Compared to RS-232, USB is faster, smaller, and simpler to use the current USB version 2.0 provides up to 480 bit/second of data transfer and will be replaced with an even faster USB 3.0 rated to 5 Gbit/second. USB also supports plug and play connectivity meaning that devices are directed by the computer's operating system and configuring automatically as soon as they are attached. USB cable can be up to 30 meters long and can also be used to bias relative relatively low power devices. These features make USB a very desirable option in the design of frequency synthesizer models since it allows instant development or simply evaluation of a synthesizer using a personal computer.

GPIB (General purpose interface bus): General purpose interface bus (GPIB) is a special interface for test and measurement applications. It was original introduced by Hewlett packers (now Agilent Technologies, Inc.) as an HP IB bus to control measurement instruments in 1975 the interface was standardized by the Institute of Electrical And Electronics Engineers (IEEE) under the IEEE–488 standards. The GPIB bus has 24 lines 8 single lines used for data transfer, three lines for handshake, five lines for bus management, and eight lines for ground returns. It allows the connection of multiple off-the-shelf instruments into a complex automated test system (ATE).

VXI (VME Extensions for Instrumentation): VXI stands for VME Extensions for Instrumentation and is an interface that was introduced in the mid 1980s as an open system platform for synthetic instrumentation. One of the principles behind synthetic instrumentation is general and VXI is particular, is to offer a cost-efficient modular approach for building complex test and measurement equipment's. It enables the emulation of various traditional bench-top instruments employed in automatic test system using reconfigurable combination of core hardware models. VXI instrument includes a chassis (also called mainframe) that contain several spaces (slots) where individual VXI models can be placed. The mainframe also contains all necessary DC power supplies and provides communication between individual components and a host controller that is usually an external computer.

VXI model such as a signal generator that delivers all necessary control and bias line, The VXI specification are governed by the VXI bus Consortium, which was founded in 1987 by a group of interested companies to define mechanical, electrical, and software features of VXI instrumentation.

PXI (PCI Extension for Instrumentation): PXI stands for PCI Extension for Instrumentation and is a future enhancement of the synthetic instrumentation concept (PCI stands for peripheral component interconnect). The PXI standard was introduced by national instruments corporation in 1997 and is currently government by the PXI systems alliance (PXI-SA). The alliance includes more than 50 companies chartered to promote the standard, ensure interoperability, and maintain PXI specifications.

Similar to VXI, a typical PXI instrument is built using PXI chassis and a number of individual models that fit into PXI slot. However, the size of the chassis and the modules is significantly similar; a typical PXI module measures approximately 4 by 6 inches in dimensions. Moreover, the host computer can be built as a PXI component and plugged in to chassis.

Therefore, a whole instrument or even and ATE system can be e completed within a single PXI frame. Another distinct advantage is higher communication speed compared to the VXI environment. It should be noted that the PXI chassis backplane uses essentially the same PCI bus used in personal computers. Thus the development and operation of PXI systems are not much different from that of standard Windows based applications. A newer PXI express standard (released in 2005) further increases the available PXI bandwidth by taking advantage of PCI express technology. User's benefits from significantly increased bandwidth ensure backward compatibility, and additional timing and synchronization features.

LXI (LAN Extensions for Instrumentation): LXI stands for LAN Extensions for Instrumentation and is another interface for test and measurement applications (LAN stands for local area network). It was introduced in 2004 by Agilent technologies,

Inc, and is currently maintained by the LXI consortium. The LXI concept offers integration advantages of modular instruments without the constraints of card-cage architectures. It is instruments into a network LXI can be used at any level of network complexity ranging from a single component and a controlling computer to complex multi instrument systems operated remotely through the internet.

The LXI standard defines three classes of instruments. The base class c incorporates a web browser via an Ethernet port as well as an interchangeable virtual instrument (I V I) driver. Class b brings synchronization capability Via the IEEE 1588 precision time protocol and also supports peer- to- peer messaging. The IEEE 1588 protocol synchronizes clock in multiple devices to ensure proper event time stamping and execution of synchronized events. Finally, Class A adds a fast hardware Trigger bus, which offers lower- latency synchronization compared to the Class B.

AXIe (Advanced TCA Extension for Instrumentation and test): AXIe (Advanced TCA Extension for Instrumentation and test) is a recent addition to the synthetic instrumentation interfaces that support both PXI and LXI standard (TCA stands for Telecommunication Computing Architecture). It is governed by the AXIe Consortium that was formed in 2009 by Agilent Technologies Inc Aeroflex Corporation and Test Evolution Corporation. AXIe addresses a wide range of ATE systems, rack-and-stack modular, bench-top, and module plug–ins. It offers higher performance par rack inch, greater scalability, more flexibility and easy integration with various platforms.

Measurement techniques and instruments

De-embedding: At RF and microwave frequency, it is important to measure S-parameter at the device termination, embedded in text fixtures for the measurement to be carried out. De-embedding is the mathematical process involving the embedding network and determining the true parameters.

Signal generator: It is an electronic device that generates repeating and non-reperating electronic signals (analog/digital).

It is used for testing, troubleshooting and repairing electronic or electro-acoustic device. They are known as test signal generators, function generators, digital pattern generators. The important aspect of the RF signal generator is:

- Frequency range
- Harmonics and spurious signals
- Power output
- Power accuracy
- Phase noise
- Accuracy (short and long term)
- Modulation format

Spectrum Analyzer: Signal analysis consisting of the sinusoid components without knowing the phase relationship. It measure frequency, power, modulation, distortion and noise.

Source: Alexander Chenakin, *Frequency Synthesizers*: *Concept to Product*, Artech House 2011

Annexure 13

Analog Vs Digital

The basic criteria differentiating between analog and digital circuits are the frequency ranges. The clock frequencies below 1 GHz in digital circuits considered higher electromagnetic signals as parasitics. This include densly packed metal traces, via,PTH etc. Above 1GHz are considered as microwave circuits which are considered as small signal and large signal dealing for the measurement of gain, harmonic, IMD and multiplication.

Present era having clock speed rises beyond 1 GHz which necessitates analysis in the microwave domain for which signal integrity analysis become important. Main parameters are ringing threshold and rise time. The other phenomena to be dealt are coupling, mismatch, discontinuity, cross talk, impedance, EMI apart from radiation. This need full wave techniques to be incorporated in complex high speed circuit analysis.

Table A13.1: Radio frequency circuit Vs Digital circuit analysis

RF circuit design	Digital circuit design
Impedance Matching	....
Frequency Response	Waveform (Eye's diagram)
Frequency Domain	Time Domain
Network Analyzer	Oscilloscope
Power Manipulation in RF Blocks	Voltage Manipulation in Digital Blocks
dBw	dBV

Digital and analogue implementations are possible, the digital devices are more versatile and commonly used. They are easily programmable, allowing large and variable division coefficients. Different logic families have different phase noise characteristics as a function of the operating frequencies. The noise in digital divider is mainly generated in the transition region where the signal cross the logic threshold in general, faster logic and larger voltage swings exhibit better phase noise performance. Analog dividers offer better phase noise characteristics. They also have the ability to work at very high frequency and are used in some critical applications.

Nyquist maximum data rate for given channel bandwidth can be represented as

$$= 2\ B \log_2 (\text{symbol})$$

where B is the channel bandwidth, symbol represent discrete symbol (BPSK–2, QPSK15–4) Shannon maximum data rate takes noise into consideration and represented as

$$= B \log_2 (1+\text{SNR})$$

SNR–signal to noise power

Annexure 14

Body Grounding of RF Devices

Body grounding of devices over the PCB plays important role in circuit performance, two reasons for grounding is safety of circuit performance and interference reduction [2]. Due to improper body grounding there are oscillations in the circuit performance, oscillations appears because of parasitic capacitances which is developed due to air gap between the device and the substrate (i.e. improper body grounding) will leads in onset of spurious, harmonics in the response which leads in reduction of power level and thereby degrading the performance of the overall circuit by not meeting the specifications. The critical aspect is the intermixing of various materials (preform, PCB substrate) which can result in the formation of inter metallic bonds and also body grounding using SMT line is not feasible for thin substrate where invariably preform to be used to avoid substrate warpage issue. The proper selection of preform plays an important role in the device attachment process. Numbers of factors are seen while selections of preform like liquidus and solidus temperature, mechanical strength, reworkability, availability, compatibility as an standby material for RF devices attachment.

The RF device is attached to the PCB following a sequence of process. The flow chart is shown in Fig. A14.1.

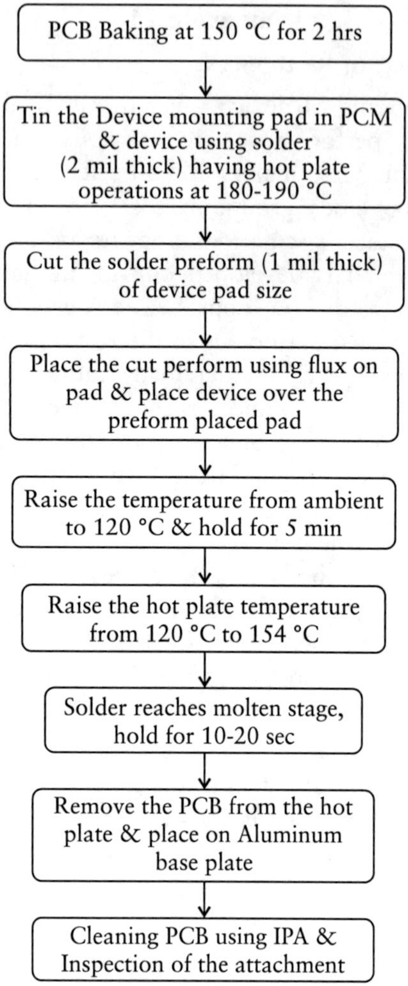

Fig. A14.1: Flow Chart of body attachment process

A modeling of device attachment over the substrate and related capacitances, inductances and resistances is shown in Fig. A14.2.

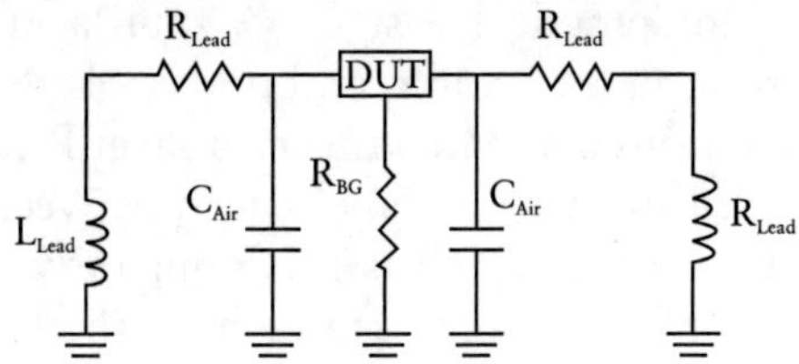

Fig. A14.2: Modelling of Body grounding

The device (DUT) leads are represented by a series combination of resistor and inductor and during the placement of the device over the substrate there is some air gap between the two surfaces which gives rise to capacitance as represented by a capacitor in parallel to the resistor, inductor combination. Due to presence of inductor and capacitor there is onset of oscillations in the circuit which degrade the performance of the circuit, in order to remove the oscillations in the performance proper body grounding is essential, and this grounding is provided by perform used in the attachment, which is represented by a resistor providing a closed loop for the flow of current and thereby eliminating the oscillations in the circuit. The tentative model and parameters can be empirically found out which can predict accurate device performance at microwave frequencies.

Annexure 15

Layout consideration in RF Boards

The main consideration in the RF board is to keep the line impedance of 50 ohms which primarily depends on the substrate height and width of the transmission line. RF PCB designs are usually designed on two- or four-layer boards. Two-layer PCB boards are designed to have component and signal routing on the top layer, while the bottom layer is designed to be predominantly ground, providing the shortest path for return currents. The ground plane should be continuous as divided layer in case of RF board increases the return current's path length. A two-layer PCB board design provides cost savings compared to a four layer PCB design but requires careful signal routing and component placement. These designs are generally limited to thicknesses of 0.8-1.0 mm, as using a greater thickness micro-strip line causes the corresponding widths for common impedances (e.g., 50 ohms) to be too large for practical designs. A four-layer PCB design provides easy routing for ground and power planes and relaxes routing considerations compared to a two-layer PCB. It provides easy decoupling of the power plane placed between the ground plane and bottom layer, which is predominantly ground. In a four-layer board it is recommended to have the layer structure as defined below. 1st Layer: Component and Signal 2nd Layer: Ground Plane 3rd Layer: Power Plane 4th Layer: Ground Plane and Signal Routing Note that in any four-layer RF PCB design a ground plane must always be below the top component and signal plane. The increased thickness (~60mils) in four-layer PCBs also provides mechanical strength to the PCB as compared to a four-layer design.

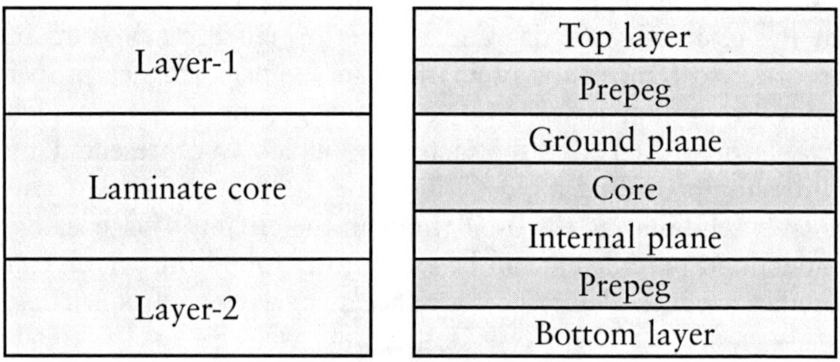

Figure A15.1: Layer stack-up for two and four layer PCB

Coupling mechanism and its mitigation

- **RF Transmission Lines:** Lines should be kept as far apart as possible, and should not be routed in close proximity for extended distances. Coupling between parallel microstrip lines will increase with decreasing separation and increasing parallel routing distance. Lines that cross on separate layers should have a ground plane keeping them apart. Signal lines that will carry high power levels should be kept away from all other line whenever possible. The grounded coplanar waveguide provides for excellent isolation between lines. It is impractical to achieve isolation better than approximately –45dB between RF lines on small PCBs.

- **High-Speed Digital Signal Lines:** These lines should be routed separately on a different layer than the RF signal lines, to prevent coupling. Digital noise (from clocks, PLLs, etc.) can couple onto RF signal lines, and these can be modulated onto RF carriers. Alternatively, in some cases digital noise can be up/down-converted.
- **VCC/Power Lines:** These should be routed on a dedicated layer. Adequate decoupling/bypass capacitors should be provided at the main VCC distribution node, as well as at VCC branches. The choice of the bypass capacitances must be made based on the overall frequency response of the RF IC, and the expected frequency distribution nature of any digital noise from clocks and PLLs. These lines should also be separated from any RF lines that will transmit large amounts of RF power.

Ground Plane consideration: The recommended practice is to use a solid (continuous) ground plane on Layer 2, assuming Layer 1 is used for the RF components and transmission lines. For strip lines and offset strip lines, ground planes above and below the center conductor are required. These planes must not be shared or assigned to signal or power nets, but must be uniquely allocated to ground. Partial ground planes on a layer, sometimes required by design constraints, must underlie all RF components and transmission lines. Ground planes must not be broken under transmission line routing. Ground vias between layers should be added liberally throughout the RF portion of the PCB. This helps prevent accrual of parasitic ground inductance due to ground-current return paths. The vias also help to prevent cross coupling from RF and other signal lines across the PCB.

Traditional PCB design imposes lot of challenges at high-frequency, high data-rate products (<100 Gb/s/channel, 5G/mm wave) primarily on signal integrity with emphasis on impedance and losses. The main concern at higher data rate and frequencies are the proximity effects and ground plane losses (20%).

The specific construction methodology can lead to various issues such as glass stop, conductive anodic filaments (CAFs), voiding, drill-induced crazing etc. and needs extensive trials to overcome for reliable performance. Further simulation for PCB thermomechanical properties, impedance losses, roughness losses to be carried out at the desired frequency.

Source: *Microwaves & RF Magazine*, 2021.

Annexure 16

Field Attenuation in Substrate

Substrate permittivity affects the conductor, dielectric and radiation loss in the microstrip line. All these mechanism are dependent on the substrate permittivity resulting in various phenomena such as surface wave, ground wave, leaky wave or radiated wave. Selection of dielectric is very important as substrate with high dielectric constant can introduce losses. It also governs the width, propagation characteristics, line length, losses and various modes of propagation. Generally dielectric loss is less compared to conductor losses in microwave region, but for lossy substrate such as silicon the effect is considerably high and affects the circuit performance considerably at higher frequencies. Dielectric loss is represented by the loss angle represented as:

$$\tan\delta = \frac{\omega\varepsilon'' + \sigma}{\omega\varepsilon'} = \tan\delta_d + \tan\delta_L \qquad \text{... (A16.1)}$$

In general case the tan δ_L is negligible and only contributing factor is from tan δ_d. The first term tan δ_d resulted from the polarization loss associated with intrinsic substrate and second term tan δ_L is defined as extrinsic loss resulted from the finite conductivity of the substrate which is applicable for silicon. Various other factors such as temperature and humidity also impact the dielectric constant value. The dielectric loss (for narrow lines) can also be represented as (in case of good dielectric)

$$\alpha_d = \frac{\sigma}{2}\sqrt{\frac{\mu_r\mu_0}{\varepsilon_r\varepsilon_0}} = 1.634\times 10^3 \frac{\sigma}{\varepsilon_r}\text{dB/cm}$$

Dielectric loss due to damping effect can be further correlated with electric susceptibility as higher susceptibility renders more heating.

The prime parameter is the attenuation for which

$$\alpha = \omega\sqrt{\frac{\mu\varepsilon}{2}}\left(1+\frac{\varepsilon''}{\varepsilon'}-1\right)\text{where } \varepsilon' = \varepsilon_r\varepsilon_0 \text{ and } \varepsilon'' = \frac{\sigma}{\omega} \quad \ldots \text{(A16.2)}$$

In case of silicon considerably electric field at the bottom surface compared to the top which can excite substrate modes and can effect the bandwidth of operation. The electric filed can be represented as:

$$\bar{E}(z) = e^{-z\alpha}\left|E\right| = e^{-z\alpha}\left|\mathrm{Re}\,E_0 e^{j(\omega t-\beta z)}\right|$$

Capacitance between the upper conductor and bottom ground plane in microstrip can be represented as:

$$C(F) = \frac{\varepsilon_r \ell}{60c\ln\left[\frac{8h}{w}+\frac{w}{4h}\right]} \text{ for } \frac{w}{h}<1$$

$$C(F) = \frac{\varepsilon_r \ell\left[\frac{w}{h}+1.393+0.667\ln\left(\frac{w}{h}+1.444\right)\right]}{120\pi.c} \text{ for } \frac{w}{h}\geq 1$$

The capacitance is directly proportional the relative permittivity of the substrate. The finite resistivity of the substrate impacts the capacitance as virtual ground is considered closer to the conductor surface. This feature makes the design of RF circuits difficult on silicon wafer. In the case of patch antenna, efficiency is enhanced with the reduction in relative permittivity and for $\varepsilon_r \sim 1$, maximum efficiency is achieved.

List of Symbols

Z_o	characteristics Impedance
Y_o	characteristics Admittance
Y_{ine}	even-mode input admittance
Y_{ino}	odd-mode input admittance
Z_{oo}	odd-mode impedance
Z_{oe}	even-mode Impedance
$\tan\delta$	loss-tangent
f_{bw}	fractional bandwidth
W	width of transmission line
f_c	cut-off frequency
f_o	center frequency
f_{res}	resonant frequency
α	attenuation constant
c	coupling coefficient
Γ	reflection constant
Γ_e	even-mode reflection constant
Γ_o	odd-mode reflection constant
$M_{i,j+1}$	Coupling/cross-coupling between resonators
ε	Permittivity
θ_e	even-mode electrical length
θ_o	odd-mode electrical length
g_i	coupling coefficient
Q_E	external quality factor

Q_U	unloaded quality factor
π	universal constant (22/7)
σ	conductivity of the medium
η	Intrinsic impedance
ω	angular frequency
c	Speed of the light in vacuum
μ	vacuum permeability
h	height of the substrate
g_m	transconductance
f_{max}	maximum frequency of operation
C_{gs}	gate-source capacitance
f_T	transition frequency
T_t	transit time
τ	minority carrier life-time
N	ideality factor
c	velocity of light
C	capacitance
h	plank's constant
e-m	electro-magnetic
k	Boltzmann constant
ω	angular velocity
σ	bulk conductivity
Z	impedance